Jarosław Stepaniuk

Rough – Granular Computing in Knowledge Discovery and Data Mining

Studies in Computational Intelligence, Volume 152

Editor-in-Chief

Prof. Janusz Kacprzyk
Systems Research Institute
Polish Academy of Sciences
ul. Newelska 6
01-447 Warsaw
Poland
E-mail: kacprzyk@ibspan.waw.pl

Jarosław Stepaniuk

Rough – Granular Computing in Knowledge Discovery and Data Mining

 Springer

Professor Jarosław Stepaniuk
Department of Computer Science
Bialystok University of Technology
Wiejska 45A, 15-351 Bialystok
Poland
Email: jstepan@wi.pb.edu.pl

ISBN 978-3-540-70800-1   e-ISBN 978-3-540-70801-8

DOI 10.1007/978-3-540-70801-8

Studies in Computational Intelligence   ISSN 1860949X

Library of Congress Control Number: 2008931009

Typeset & Cover Design: Scientific Publishing Services Pvt. Ltd., Chennai, India.

Printed in acid-free paper
9 8 7 6 5 4 3 2 1
springer.com

To
Elżbieta and Anna

Foreword

> If controversies were to arise, there would be no more need of
> disputation between two philosophers than between two
> accountants. For it would suffice to take their pencils in their hands,
> and say to each other: 'Let us calculate'.
> **Gottfried Wilhelm Leibniz (1646–1716)**
> **Dissertio de Arte Combinatoria (Leipzig, 1666)**

Gottfried Wilhelm Leibniz, one of the greatest mathematicians, discussed calculi
of thoughts. Only much later, did it become evident that new tools are necessary
for developing such calculi, e.g., due to the necessity of reasoning under uncertainty about objects and (vague) concepts. Fuzzy set theory (Lotfi A. Zadeh,
1965) and rough set theory (Zdzisław Pawlak, 1982) represent two different approaches to vagueness. Fuzzy set theory addresses gradualness of knowledge,
expressed by the fuzzy membership, whereas rough set theory addresses granularity of knowledge, expressed by the indiscernibility relation. Granular computing (Zadeh, 1973, 1998) is currently regarded as a unified framework for theories,
methodologies and techniques for modeling calculi of thoughts, based on objects
called granules.

The book "Rough–Granular Computing in Knowledge Discovery and Data
Mining" written by Professor Jaroslaw Stepaniuk is dedicated to methods based
on a combination of the following three closely related and rapidly growing areas: granular computing, rough sets, and knowledge discovery and data mining
(KDD). In the book, the KDD foundations based on the rough set approach
and granular computing are discussed together with illustrative applications. In
searching for relevant patterns or in inducing (constructing) classifiers in KDD,
different kinds of granules are modeled. In this modeling process, granules called
approximation spaces play a special rule. Approximation spaces are defined by
neighborhoods of objects and measures between sets of objects. In the book,
the author underlines the importance of approximation spaces in searching for

relevant patterns and other granules on different levels of modeling for compound concept approximations. Calculi on such granules are used for modeling computations on granules in searching for target (sub) optimal granules and their interactions on different levels of hierarchical modeling. The methods based on the combination of granular computing, the rough and fuzzy set approaches allow for an efficient construction of the high quality approximation of compound concepts.

The book "Rough–Granular Computing in Knowledge Discovery and Data Mining" is an important contribution to the literature. The author and the publisher, Springer, deserve our thanks and congratulations.

March 30, 2008   Andrzej Skowron
Warsaw, Poland

Preface

The purpose of computing is insight, not numbers.
Richard Wesley Hamming (1915–1998)
Art of Doing Science and Engineering: Learning to Learn

Lotfi Zadeh has pioneered a research area known as computing with words. The objective of this research is to build intelligent systems that perform computations on words rather than on numbers. The main notion of this approach is related to information granulation. Information granules are understood as clumps of objects that are drawn together by similarity, indiscernibility or functionality. Granular computing may be regarded as a unified framework for theories, methodologies and techniques that make use of information granules in the process of problem solving.

Zdziaław Pawlak has pioneered a research area known as rough sets. A lot of interesting results were obtained in this area. We only mention that, recently, the seventh volume of an international journal, Transactions on Rough Sets was published. This journal, a subline in the Springer series Lecture Notes in Computer Science, is devoted to the entire spectrum of rough set related issues, starting from foundations of rough sets to relations between rough sets and knowledge discovery in databases and data mining.

This monograph is dedicated to a newly emerging approach to knowledge discovery and data mining, called rough–granular computing. The emerging concept of rough–granular computing represents a move towards intelligent systems. While inheriting various positive characteristics of the parent subjects of rough sets, clustering, fuzzy sets, etc., it is hoped that the new area will overcome many of the limitations of its forebears. A principal aim of this monograph is to stimulate an exploration of ways in which progress in data mining can be enhanced through integration with rough sets and granular computing.

The monograph has been very much enriched thanks to foreword written by Professor Andrzej Skowron. I also would like to thank him for his encouragement and advice.

I am very thankful to Professor Janusz Kacprzyk who supported the idea of this book.

The research was supported by the grants N N516 069235 and N N516 368334 from Ministry of Science and Higher Education of the Republic of Poland and by the grant Innovative Economy Operational Programme 2007-2013 (Priority Axis 1. Research and development of new technologies) managed by Ministry of Regional Development of the Republic of Poland.

April 2008   Jarosław Stepaniuk
Białystok, Poland

Contents

Part IV: Conclusions, Bibliography and Further Readings

1 Introduction

The amount of electronic data available is growing very fast and this explosive growth in databases has generated a need for new techniques and tools that can intelligently and automatically extract implicit, previously unknown, hidden and potentially useful information and knowledge from these data. These tools and techniques are the subject of the fields of knowledge discovery in databases and data mining.

In [218] ten most important problems in data mining research were identified. We summarize ten problems below:

1. **Developing a unifying theory of data mining.** The current state of the art of data mining research seems too ad-hoc. Many techniques are designed for individual problems, such as classification of objects or clustering, but there is no unifying theory. However, a theoretical framework that unifies different data mining tasks including clustering, classification, association rules would help the field and provide a basis for future research.
2. **Scaling up for high dimensional data and high speed data streams.** One challenge is how to design classifiers to handle ultra-high dimensional classification problems. There is a strong need now to build useful classifiers with hundreds of millions of attributes, for applications such as text mining and drug safety analysis. Such problems often begin with tens of thousands of attributes and also with interactions between the attributes, so the number of discovered new attributes gets huge quickly. One important problem is mining data streams in extremely large databases.
3. **Mining sequence data and time series data.** Sequential and time series data mining remains an important problem. Despite progress in other related fields, how to efficiently cluster, classify and predict the trends of these data is still an important open topic. Examples of these applications include the predictions of financial time series and seismic time series. In [60] is proposed approach to evaluating perception that provides a basis for optimizing various tasks related to discovery of compound granules representing process models, their interaction, or approximation of trajectories of

J. Stepaniuk: Rough - Gran. Comput. in Knowl. Dis. & Data Min., SCI 152, pp. 1–9, 2008.
springerlink.com   © Springer-Verlag Berlin Heidelberg 2008

discovered models of processes. In [62] and [63] is proposed a new approach to the linguistic summarization of time series data.

4. **Mining complex knowledge from complex data.** One important type of complex knowledge can occur when mining data from multiple relations. In most domains, the objects of interest are not independent of each other, and are not of a single type. We need data mining systems that can soundly mine the rich structure of relations among objects, such as interlinked Web pages, social networks, metabolic networks in the cell, etc. In particular, one important area is to incorporate background knowledge into data mining.

5. **Data mining in a network setting.** Network mining problems pose a key challenge. Network links are increasing in speed. To be able to detect anomalies (e.g. sudden traffic spikes due to a denial of service attack or catastrophic event), service providers will need to be able to capture IP packets at high link speeds and also analyze massive amounts of data each day. One will need highly scalable solutions for this problem.

6. **Distributed data mining and mining multi-agent data.** The problem of distributed data mining is very important in network problems. In a distributed environment the problem is to discover patterns in the global data seen at all the different places. There could be different models of distributed data mining, but the goal obviously would be to minimize the amount of data shipped between the various sites essentially, to reduce the communication overhead. In distributed mining, one problem is how to mine across multiple heterogeneous data sources: multi-database and multi-relational mining.

7. **Data mining for biological and environmental problems.** Many researchers believe that mining biological data continues to be an extremely important problem, both for data mining research and for biomedical sciences.

8. **Data mining process-related problems.** Important topics exist in improving data-mining tools and processes through automation. Specific issues include how to automate the composition of data mining operations and building a methodology into data mining systems to help users avoid many data mining mistakes. There is also a need for the development of a theory behind interactive exploration of complex data.

9. **Security, privacy and data integrity.** Related to the data integrity assessment issue, the two most significant challenges are: develop efficient algorithms for comparing the knowledge contents of the two (before and after) versions of the data, and develop algorithms for estimating the impact that certain modifications of the data have on the statistical significance of individual patterns obtainable by broad classes of data mining algorithms.

10. **Dealing with non-static, unbalanced and cost-sensitive data.** An important issue is that the learned pattern should incorporate time because data is not static and is constantly changing in many domains. Another related issue is how to deal with unbalanced and cost-sensitive data, a major challenge in data mining research.

In this book we discuss selected rough-granular computing solutions to some above mentioned data mining problems.

Granular computing is inspired by Zadeh's definition of information granule: "Information granule is a clump of objects drawn together by indiscernibility, similarity or functionality." We start from elementary granules based on indiscernibility classes (as in the standard rough set model) and tolerance classes (as in the tolerance rough set model) and investigate complex information granules. Granular computing (GC, in short) may be regarded as a unified framework for theories and methodologies that make use of granules in the process of problem solving. Granulation leads to information compression. Therefore computing with granules, rather than objects provides gain in computation time, thereby making the role of granular computing significant in knowledge discovery and data mining.

Rough-granular computing (RGC, in short) is defined as granular computing based on the rough set approach.

Knowledge Discovery in Databases (KDD, for short) has been defined as "the nontrivial extraction of implicit, previously unknown, and potentially useful information from data" [21, 34]. Among others, it uses machine learning, rough sets, statistical and visualization techniques to discover and present knowledge in a form easily comprehensible to humans. Knowledge discovery is a process which helps to make sense of data in a more readable and applicable form. It usually involves at least one of two different goals: description and classification (prediction). Description focuses on finding user-interpretable patterns describing the data. Classification (prediction) involves using some attributes in the data table to predict values (future values) of other attributes (see. e.g. [71]).

The theory of rough sets provides a powerful foundation for discovery of important regularities in data and for objects classification. In recent years numerous successful applications of rough set methods for real-life data have been developed (see e.g. [103, 106, 108, 109, 110, 123, 124]).

We will now describe in some detail main contributions of this book.

Rough sets: classification of objects by means of attributes. Rough set approach has been used in a lot of applications aimed to description of concepts. In most cases only approximate descriptions of concepts can be constructed because of incomplete information about them. Let us consider a typical example for classical rough set approach when concepts are described by positive and negative examples. In such situations it is not always possible describe concepts exactly, since some positive and negative examples of the concepts being described inherently can not be distinguished one from another. Rough set theory was proposed [106] as a new approach to vague concept description from incomplete data. The rough set approach to processing of incomplete data is based on the lower and the upper approximation. The rough set is defined as the pair of two crisp sets corresponding to approximations. If both approximations of a given subset of the universe are exactly the same, then one can say that the subset mentioned above is definable with respect to available information. Otherwise, one can consider it as roughly definable. Suppose we are given a finite non-empty set U of objects, called the universe. Each object of U is characterized by a description constructed, for example from a set of attribute values. In standard rough set approach [106] introduced by Pawlak an equivalence relation (reflexive, symmetric and transitive relation) on the universe of objects is defined from equivalence relations on the attribute values. In particular, this equivalence relation is constructed assuming the existence of the equality relation on attribute values. Two different objects are indiscernible in view of available information, because with these objects the same information can be associated. Thus, information associated with objects from the universe generates an indiscernibility relation in this universe. In the standard rough set model the lower approximation of any subset $X \subseteq U$ is defined as the union of all equivalence classes fully included in X. On the other hand the upper approximation of X is defined as the union of all equivalence classes with a non-empty intersection with X.

In real data sets usually there is some noise, caused for example from imprecise measurements or mistakes made during collecting data. In such situations the notions of "full inclusion" and "non-empty intersection" used in approximations definition are too restrictive. Some extensions in this direction have been proposed by Ziarko in the variable precision rough set model [229].

The indiscernibility relation can be also employed in order to define not only approximations of sets but also approximations of relations [29, 43, 101, 105, 138, 141, 177, 185]. Investigations on relation approximation are well motivated both from theoretical and practical points of view. Let us bring two examples. The equality approximation is fundamental for a generalization of the rough set approach based on a similarity relation approximating the equality relation in the value sets of attributes. Rough set methods in control processes require function approximation.

However, the classical rough set approach is based on the indiscernibility relation defined by means of the equality relations in different sets of attribute values. In many applications instead of these equalities some similarity (tolerance) relations are given only. This observation has stimulated some researchers to generalize the rough set approach to deal with such cases, i.e., to consider similarity (tolerance) classes instead of the equivalence classes as elementary definable sets. There is one more basic notion to be considered, namely the rough inclusion of concepts. This kind of inclusion should be considered instead of the exact set equality because of incomplete information about the concepts. The two notions mentioned above, namely the generalization of equivalence classes to similarity classes (or in more general cases to some neighborhoods) and the equality to rough inclusion have lead to a generalization of classical approximation spaces defined by the universe of objects together with the indiscernibility relation being an equivalence relation. We discuss applications of such approximation spaces for solution of some basic problems related to concept descriptions.

One of the problems we are interested in is the following: given a subset $X \subseteq U$ or a relation $R \subseteq U \times U$, define X or R in terms of the available information. We discuss an approach based on generalized approximation spaces introduced and investigated in [141, 145]. We combine in one model not only some extension of an indiscernibility relation but also some extension of the standard inclusion used in definitions of approximations in the standard rough set model. Our approach allows to unify different cases considered for example in [106, 229].

There are several modifications of the original approximation space definition [106]. The first one concerns the so called uncertainty function. Information about an object, say x is represented for example by its attribute value vector. Let us denote the set of all objects with similar (to attribute value vector of x) value vectors by $I(x)$. In the standard rough set approach [106] all objects with the same value vector create the indiscernibility class. The relation $y \in I(x)$ is in this case an equivalence relation. The second modification of the approximation space definition introduces a generalization of the rough membership function [107]. We assume that to answer a question whether an object x belongs to an object set X we have to answer a question whether $I(x)$ is in some sense included in X.

Approximation spaces based on uncertainty functions and rough inclusions were also investigated in [142, 145, 158, 186, 189]. Some comparison of standard approximation spaces [106] and the above mentioned approach in approximation of concepts was presented in [42].

Reducts. We start with short history about top data mining algorithms [217]. Finding reduct algorithm [106] was in the nominations for top ten data mining algorithms. ACM KDD Innovation Award and IEEE ICDM Research Contributions Award winners nominate up to 10 best-known algorithms in data mining. Each nomination was verified for its citations on Google Scholar. Finding reduct algorithm was in the 18 identified candidates for top ten algorithms in data mining (for more details see [217]).

The ability to discern between perceived objects is important for constructing many entities like reducts, decision rules or decision algorithms. In the classical rough set approach the discernibility relation is defined as the complement of the indiscernibility relation. However, this is, in general, not the case for the generalized approximation spaces. The idea of Boolean reasoning is based on construction for a given problem P a corresponding Boolean function g_P with the following property: the solutions for the problem P can be decoded from prime implicants of the Boolean function g_P. Let us mention that to solve real-life problems it is necessary to deal with Boolean functions having large number of variables. A successful methodology based on the discernibility of objects and Boolean reasoning has been developed for computing of many important for applications entities like reducts and their approximations, decision rules, association rules, discretization of real value attributes, symbolic value grouping, searching for new features defined by oblique hyperplanes or higher order surfaces, pattern extraction from data as well as conflict resolution or negotiation (for references see the papers and bibliography in [103, 123, 124]). Most of the problems related to generation of the above mentioned entities are NP-complete or NP-hard. However, it was possible to develop efficient heuristics returning suboptimal solutions of the problems. The results of experiments on many data sets are very promising. They show very good quality of solutions generated by the heuristics in comparison with other methods reported in literature (e.g. with respect to the classification quality of unseen objects). Moreover, they are very efficient from the point of view of time necessary for computing of the solution. It is important to note that the methodology allows to construct heuristics having a very important *approximation property* which can be formulated as follows: expressions generated by heuristics (i.e., implicants) *close* to prime implicants define approximate solutions for the problem. The detailed comparison of rough set classification methods based on combination of Boolean reasoning and approximate Boolean reasoning methodology and discernibility notion with other classification methods one can find in books [103, 123, 124] and in paper [95].

Methods of Boolean reasoning for reducts and rule computation in standard and tolerance rough set model were also investigated in [95, 145, 186, 189].

Knowledge discovery in medical data. Developed so far rough set methods have shown to be very useful in many real life applications. Rough set based software systems, such as RSES [15], ROSETTA [100], LERS [44], [45] and Rough Family [166] have been applied to KDD problems. The patterns discovered by the above systems are expressed in attribute-value languages. There are numerous areas of successful applications of rough set software systems (for reviews see [104]).

We present applications of rough set and clustering methods to knowledge discovery in real life medical data set [187, 189, 197]. We consider four sub-tasks:

- identification of the most relevant condition attributes,
- application of nearest neighbor algorithms for rough set based reduced data,
- discovery of decision rules characterizing the dependency between values of condition attributes and decision attribute,
- information granulation using clustering.

The nearest neighbor paradigm provides an effective approach to classification and is one of the top ten algorithms in data mining [217]. The k-nearest neighbor (kNN) classification finds a group of k objects in the training set that are closest to the test object, and bases the assignment of a decision class on the predominance of a particular class in this neighborhood. There are three key elements of this approach: a set of labeled objects, e.g., a decision table, a distance or similarity metric to compute distance between objects, and the value of k, the number of nearest neighbors. To classify new object, the distance of this object to the labeled objects is computed, its k-nearest neighbors are identified, and the decision class of these nearest neighbors are then used to determine the decision class the object.

A major advantage of nearest neighbor algorithms is that they are non-parametric, with no assumptions imposed on the data other than the existence of a metric. However, nearest neighbor paradigm is especially susceptible to the presence of irrelevant attributes. We use the rough set approach for selection of the most relevant attributes within the diabetes data set. Next nearest neighbor algorithms are applied with respect to reduced set of attributes.

The medical information system is presented at the end of the paper [189].

Mining knowledge from complex data. In learning approximations of complex concepts there is a need to choose a description language. This choice may limit the domains to which a given algorithm can be applied. There are at least two basic types of objects: structured and unstructured. An unstructured object is usually described by attribute-value pairs. For objects having an internal structure first order logic language is often used. In the book we investigate both types of objects. In the former case we use the propositional language with atomic formulas being selectors (i.e. pairs *attribute=value*), in the latter case we consider the first order language.

Attribute-value languages have the expressive power of propositional logic. These languages sometimes do not allow for proper representation of complex structured objects and relations among objects or their components. The background knowledge that can be used in the discovery process is of a restricted form and other relations from the database cannot be used in the discovery process. Using first-order logic (or FOL for short) has some advantages over propositional logic. First order logic provides a uniform and very expressive means of representation. The background knowledge and the examples, as well as the induced patterns, can all be represented as formulas in a first order language. Unlike propositional learning systems, the first order approaches do not require that the relevant data be composed into single relation but, rather can take into account data, which is organized in several database relations with various connections existing among them. First order logic can face problems which cannot be reduced to propositional logics, such as recurrent structures. On the other hand, even if a problem can be reduced to propositional logics, the solutions found in FOL are more readable and simpler than the corresponding ones in propositional logics.

We consider some directions in applications of rough set methods to discovery of interesting patterns expressed in a first order language. The first direction is based on translation of data represented in first-order language to decision table [106] format and next on processing by using rough set methods based on the notion of a reduct. Our approach is based on the iterative checking whether a new attribute adds to the information [198]. The second direction concerns reduction of the size of the data in first-order language and is related to results described in [86, 198]. The discovery process is performed only on well-chosen portions of data which correspond to approximations in the rough set theory. Our approach is based on iteration of approximation operators [198]. The third approach to mining knowledge from complex data is based on the RSRL (**R**ough **S**et **R**elational **L**earning) algorithm [194, 195]. Rough set methods in multi-relational knowledge discovery were also investigated in [191, 192].

Complex concept approximations. One of the rapidly developing areas in computer science is now granular computing (see e.g. [112, 113, 227, 228]). Several approaches have been proposed toward formalization of the Computing with Words paradigm formulated by Lotfi Zadeh. Information granulation is a very natural concept, and appears (under different names) in many methods related to e.g. data compression, divide and conquer, interval computations, clustering, fuzzy sets, neighborhood systems, and rough sets among others. Notions of a granule and granule similarity (inclusion or closeness) are also very natural in knowledge discovery.

We present a rough set approach for granular computing. The presented approach seems to be important for knowledge discovery in distributed environment and for extracting generalized patterns from data (see problem "Distributed data mining and mining multi-agent data" [218]). We discuss the basic notions related to information granulation, namely the information granule syntax and semantics as well as the inclusion and closeness (similarity) relations of granules. We discuss some problems of generalized pattern extraction from data assuming knowledge is represented in the form of information granules. We emphasize the importance of information granule application to extract robust patterns from data. We also propose to use complex information granules to extract patterns from data in distributed environment. These patterns can be treated as a generalization of association rules.

Information granules synthesis in knowledge discovery was also investigated in [149, 150, 190].

One of the main goals of the book is to illustrate different important issues of granular computing by examples based on the rough set approach. In Chapters 2, 4, and 5 are presented methods for defining granules on different levels of modeling, e.g., elementary granules, approximation spaces, classifiers or clusters. Moreover, approximations of granules defined by decision classes by granules defined by conditional attributes are used as examples of some other more compound granules. In Chapter 2, are also presented examples of quality measures defined on granules and the optimization measures used in searching for the target granules. The description size of granules is another important issue of GC. Different kinds of reducts discussed in Chapter 3 can be treated as illustrative examples related to this issue. Granules are constructed under uncertainty from samples of some more elementary granules. Hence, methods for inducing granules with relevant properties on their extensions play important role in GC. Strategies for inducing classifiers and clusters discussed in Chapters 4 and 5 are examples of such methods. Among such methods are methods for fusion of the existing granules for obtaining more general relevant granules. This also requires developing of the quality measures used for defining the qualities of more compound granules from the qualities of less compound ones. Examples of granules used in data mining from complex data are included in Chapter 7. A general discussion on granular computing in searching for the complex concept approximations is presented in Chapter 8.

The organization of the book is as follows.

In Chapter 2 we discuss standard and extended rough set models.

In Chapter 3 we discuss reducts and representatives in standard and tolerance rough set models.

In Chapter 4 we investigate decision rules generation in standard and tolerance rough set models. We discuss also different quantitative measures associated with rules.

In Chapter 5 we discuss selected clustering algorithms. We also present some quality measures of information granulation.

In Chapter 6 we investigate knowledge discovery in real life medical data table.

In Chapter 7 we apply rough set concepts to mining knowledge from complex data.

In Chapter 8 we discuss information granules in complex concepts approximation.

At the end of the book, we give a literature in two parts:

- bibliography (cited in the book),
- further readings (books and reviews uncited in the book but of interest for further information).

Part I

Rough Set Methodology

2 Rough Sets

Rough set theory due to Zdzisław Pawlak (1926–2006) [106, 108, 109, 110], is a mathematical approach to imperfect knowledge. The problem of imperfect knowledge has been tackled for a long time by philosophers, logicians and mathematicians. Recently it has become a crucial issue for computer scientists as well, particularly in the area of computational intelligence [129], [99]. There are many approaches to the problem of how to understand and manipulate imperfect knowledge. The most successful one is, no doubt, the fuzzy set theory proposed by Lotfi A. Zadeh [226]. Rough set theory presents still another attempt to solve this problem. It is based on an assumption that objects are perceived by partial information about them. Due to this some objects can be indiscernible. Indiscernible objects form elementary granules. From this fact it follows that some sets can not be exactly described by available information about objects. They are rough not crisp. Any rough set is characterized by its (lower and upper) approximations.

One of the consequences of perceiving objects using only available information about them is that for some objects one cannot decide if they belong to a given set or not. However, one can estimate the degree to which objects belong to sets. This is another crucial observation in building foundations for approximate reasoning. In dealing with imperfect knowledge one can only characterize satisfiability of relations between objects to a degree, not precisely. Among relations on objects the rough inclusion relation, which describes to what degree objects are parts of other objects, plays a special role. A rough mereological approach (see, e.g., [104, 122, 154]) is an extension of the Leśniewski mereology [77] and is based on the relation *to be a part to a degree*. It will be interesting to note here that Jan Łukasiewicz was the first who started to investigate the inclusion to a degree of concepts in his discussion on relationships between probability and logical calculi [79].

In the rough set approach, we are searching for data models using the minimal length principles. Searching for models with small size is performed by means of many different kinds of reducts, i.e., minimal sets of attributes preserving some constraints (see Chapter 3).

J. Stepaniuk: Rough - Gran. Comput. in Knowl. Dis. & Data Min., SCI 152, pp. 13–41, 2008.
springerlink.com

One of the very successful techniques for rough set methods is Boolean reasoning. The idea of Boolean reasoning is based on constructing for a given problem P a corresponding Boolean function g_P with the following property: the solutions for the problem P can be decoded from prime implicants of the Boolean function g_P (see Figure 3.1). It is worth to mention that to solve real-life problems it is necessary to deal with Boolean functions having a large number of variables.

A successful methodology based on the discernibility of objects and Boolean reasoning has been developed in rough set theory for the computing of many key constructs like reducts and their approximations, decision rules, association rules, discretization of real value attributes, symbolic value grouping, searching for new features defined by oblique hyperplanes or higher order surfaces, pattern extraction from data as well as conflict resolution or negotiation (see, e.g., [95, 134]). Most of the problems involving the computation of these entities are NP-complete or NP-hard. However, we have been successful in developing efficient heuristics yielding sub-optimal solutions for these problems. The results of experiments on many data sets are very promising. They show very good quality of solutions generated by the heuristics in comparison with other methods reported in literature (e.g., with respect to the classification quality of unseen objects). Moreover, they are very time-efficient. It is important to note that the methodology makes it possible to construct heuristics having a very important approximation property. Namely, expressions generated by heuristics (i.e., implicants) close to prime implicants define approximate solutions for the problem (see, e.g., [15]).

Standard rough set model is based on equivalence relations (see Section 2.1.3). The notion of tolerance relation (see Section 2.1.4) is a basis for tolerance rough set model. In this chapter we discuss basic concepts for standard and tolerance rough set models. We investigate an idea to turn the equivalence into tolerance relation, for more expressive modeling of the lower and upper approximations of a crisp set.

The chapter is organized as follows. In Section 2.1 we recall basic concepts of equivalence relations and tolerance relations. In Section 2.2 the notion of information system is recalled. In Section 2.3 properties of approximations in generalized approximation spaces are discussed. In Section 2.4 approximations of relations are investigated. In Section 2.5 the notion of function approximation is discussed. In Section 2.6 we discuss in detail some quality measures of approximation spaces. In Section 2.7 we discuss conventional and evolutionary strategies for learning approximation space from data. In Section 2.8 we give general remarks about rough sets in concept approximation.

2.1 Preliminary Notions

Based on the literature, in this section we discuss basic concepts of equivalence relations and tolerance relations.

2.1.1 Sets

The notion of a set is a basic one of mathematics. Most mathematical structures refer to it. The area of mathematics that deals with collections of objects, their properties and operations is called set theory. The creation of set theory is due to German mathematician Georg Cantor (1845–1918).

The fact that an element x belongs to a set X is denoted by $x \in X$ and the notation $x \notin Y$ denotes that the element x is not a member of the set Y.

For the finite set X, cardinality, denoted by $card(X)$, is the number of set elements. For example, $card(\{1, a, 2\}) = 3$.

A set X is a subset of set Y ($X \subseteq Y$) if and only if every element of X is also member of set Y.

The power set of a given set X (denoted by $P(X)$) is the collection of all possible subsets of X. For example, the power set of the set $X = \{1, a, 2\}$ is $P(X) = \{\emptyset, \{1\}, \{a\}, \{2\}, \{1, a\}, \{1, 2\}, \{a, 2\}, \{1, a, 2\}\}$.

Let $X = \{x_1, x_2, \ldots\}$ and $Y = \{y_1, y_2, \ldots\}$. The Cartesian product of two sets X and Y, denoted by $X \times Y$, is the set of all ordered pairs (x, y) of elements $x \in X$ and $y \in Y$.

Given a non-empty set U, any subset $R \subseteq U \times U$ is called a binary relation in U.

2.1.2 Properties of Relations

We consider here certain properties of binary relations.

Definition 2.1. Reflexivity. *Given a non-empty set U and a binary relation $R \subseteq U \times U$, R is reflexive if and only if all the ordered pairs of the form (x, x) are in R for every $x \in U$.*

A relation which fails to be reflexive is called nonreflexive. We always consider relations in some set and a relation (considered as a set of ordered pairs) can have different properties in different sets. For example, the relation $R = \{(1, 1), (2, 2)\}$ is reflexive in the set $U_1 = \{1, 2\}$ and nonreflexive in $U_2 = \{1, 2, 3\}$ since it lacks the pair $(3, 3)$.

Definition 2.2. Symmetry. *A relation $R \subseteq U \times U$, is symmetric if and only if for every ordered pair $(x, y) \in U \times U$ if (x, y) is in R, then the pair (y, x) is also in R.*

If for some $(x, y) \in R$, the pair (y, x) is not in R, then R is nonsymmetric.

Definition 2.3. Transitivity. *A relation $R \subseteq U \times U$, is transitive if and only if for all $x, y, z \in U$ if $(x, y) \in R$ and $(y, z) \in R$, then the pair (x, z) is in R.*

Using properties of relations we can consider some important classes of relations, namely, equivalence relations and tolerance relations.

2.1.3   Equivalence Relations

Definition 2.4. *An equivalence relation is a relation which is reflexive, symmetric and transitive.*

For every equivalence relation there is a natural way to divide the set on which it is defined into mutually exclusive (disjoint) subsets which are called equivalence classes. We write $[x]_R$ for the set of all y such that $(x, y) \in R$. Thus, when $R \subseteq U \times U$ is an equivalence relation, $[x]_R$ is the equivalence class which contains x. The set $U/R = \{[x]_R : x \in U\}$ is called a quotient set of the set U by the equivalence R. U/R is a subset of $P(U)$ (the set of all subsets of U).

The relations "has the same hair color as" or "is the same age as" in the set of people are equivalence relations. The equivalence classes under the relation "has the same hair color as" are the set of blond people, the set of red-haired people, etc.

Definition 2.5. Partition. *Given a non-empty set U, a partition of U is a collection of non-empty subsets of U such that*

1. *for any two distinct subsets $X \subseteq U$ and $Y \subseteq U$, $X \cap Y = \emptyset$,*
2. *the union of all the subsets in collection equals U.*

Let us consider the set $U = \{1, a, 2\}$. The set $\{\{1, 2\}, \{a\}\}$ is a partition of the set U. However, the set $\{\{1, 2\}, \{1, a\}\}$ is not a partition, because its members are not disjoint.

The subsets of U that are members of a partition of U are called cells of that partition. There is a close correspondence between partitions and equivalence relations. Given a partition of set U, the relation $R = \{(x, y) \in U \times U : x$ and y are in the same cell of the partition of $U\}$ is an equivalence relation in U. Conversely, given an equivalence relation R in U, there exists a partition of U in which x and y are in the same cell if and only if $(x, y) \in R$.

2.1.4   Tolerance Relations

Definition 2.6. *A relation $R \subseteq U \times U$ is called a tolerance relation if and only if it is reflexive and symmetric.*

So tolerance is weaker than equivalence; it does not need to be transitive. The notion of tolerance relation is an explication of similarity or closeness. Relations "neighbor of", "friend of" can be considered as examples if we hold that every person is a neighbor and a friend to him(her)self. As analogs of equivalence classes and partitions, here we have tolerance classes and coverings. A set $X \subseteq U$ is called a tolerance preclass if it holds that for all $x, y \in X$, x and y are tolerant, i.e. $(x, y) \in R$. A maximum preclass is called a tolerance class. So two tolerance classes can have common elements.

Definition 2.7. Covering. *Given a non-empty set U, a collection (set) P of non-empty subsets of U such that $\bigcup_{X \in P} = U$ is called a covering of U.*

Given a tolerance relation in U, the collection of its tolerance classes forms a covering of U. Every partition is a covering but not every covering is a partition. For cxample, the set $\{\{1, 2\}, \{1, a\}\}$ is a covering of the set $U = \{1, a, 2\}$.

2.2 Information Systems

In his seminal book, Pawlak [106] introduced the notion of information system, also determined as knowledge representation system. In this section, we recall some basic definitions.

Let U denote a finite non-empty set of objects, to be called the universe. Further, let A denote a finite non-empty set of attributes. Every attribute $a \in A$ is a function

$$a : U \rightarrow V_a,$$

where V_a is the set of all possible values of a, to be called the domain of a. In the sequel, $a(x)$, $a \in A$ and $x \in U$, denotes the value of attribute a for object x.

Definition 2.8. *A pair $IS = (U, A)$ is an information system.*

Usually, the specification of an information system can be presented in tabular form.

Each subset of attributes $B \subseteq A$ determines a binary $B - indiscernibility$ relation $IND(B)$ consisting of pairs of objects indiscernible with respect to attributes from B. Thus, $IND(B) = \{(x, y) \in U \times U : \forall_{a \in B} a(x) = a(y)\}$. $IND(B)$ is an equivalence relation and determines a partition of U, which is denoted by

Table 2.1. An Information System

U	a_1	a_2	a_3
x_1	f	early_school	short
x_2	m	preschool	short
x_3	f	adolescence	medium
x_4	m	preschool	short
x_5	m	early_school	short
x_6	m	adolescence	short
x_7	f	adolescence	long
x_8	f	preschool	medium
x_9	m	adolescence	medium

Table 2.2. Partitions Defined by Indiscernibility Relations

$IND(\bullet)$	Partition $U/IND(\bullet)$
$\{a_1\}$	$\{\{x_1, x_3, x_7, x_8\}, \{x_2, x_4, x_5, x_6, x_9\}\}$
$\{a_1, a_2\}$	$\{\{x_1\}, \{x_2, x_4\}, \{x_3, x_7\}, \{x_5\}, \{x_6, x_9\}, \{x_8\}\}$
$\{a_1, a_2, a_3\}$	$\{\{x_1\}, \{x_2, x_4\}, \{x_3\}, \{x_5\}, \{x_6\}, \{x_7\}, \{x_8\}, \{x_9\}\}$

$U/IND(B)$. The set of objects indiscernible with an object $x \in U$ with respect to B in IS is denoted by $I_B(x)$ and is called $B - indiscernibility$ class. Thus, $I_B(x) = \{y \in U : (x, y) \in IND(B)\}$ and $U/IND(B) = \{I_B(x) : x \in U\}$.

Definition 2.9. *A pair $AS_B = (U, IND(B))$ is a standard approximation space for the information system $IS = (U, A)$, where $B \subseteq A$.*

Example 2.10. The information system was adopted from the paper [189]. This is the real life medical data set (see Chapter 6 for more details). For simplicity of presentation we only consider part of this data set, namely $IS = (U, A)$, where $U = \{x_1, \ldots x_9\}$ and $A = \{a_1, a_2, a_3\}$. The attribute a_1 means sex, the attribute a_2 means age of disease diagnosis and the attribute a_3 means disease duration (see Table 2.1). We obtain $V_{a_1} = \{f, m\}$, $V_{a_2} = \{preschool, early\_school, adolescence\}$ and $V_{a_3} = \{short, medium, long\}$.

Some examples of partitions defined by indiscernibility relations for information system in Table 2.1 are given in Table 2.2.

In his book, Pawlak [106] gives also a formal definition of a decision table. An information system with distinguished conditional attributes and decision attribute is called a decision table.

Definition 2.11. *A tuple $DT = (U, A \cup \{d\})$, where $d \notin A$ is a decision table.*

We will also use notation (U, A, d) for the decision table DT.

2.3 Approximation Spaces

In this section, we recall the definition of an approximation space from [141], [145], [186], [189]. Approximation spaces can be treated as granules used for concept approximation. They are some special parameterized relational structures. Tuning of parameters makes it possible to search for relevant approximation spaces relative to given concepts.

For every non-empty set U, let $P(U)$ denote the set of all subsets of U (the power set of U).

Definition 2.12. *A parameterized approximation space is a system $AS_{\#,\$} = (U, I_\#, \nu_\$)$, where*

- *U is a non-empty set of objects,*
- *$I_\# : U \to P(U)$ is an uncertainty function,*
- *$\nu_\$: P(U) \times P(U) \to [0, 1]$ is a rough inclusion function,*

and $\#, \$$ denote vectors of parameters (the indexes $\#, \$$ will be omitted if it does not lead to misunderstanding).

An idea of approximation space is depicted on Figure 2.1.

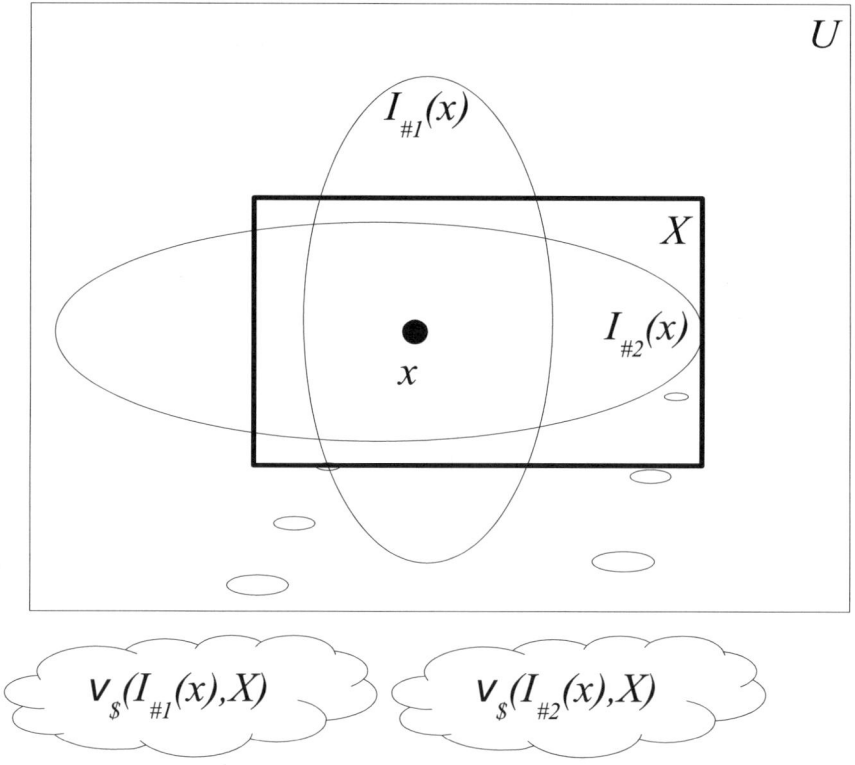

Fig. 2.1. Parameterized Approximation Space

2.3.1   Uncertainty Function

The uncertainty function defines for every object x, a set of objects described similarly to x. The set $I(x)$ is called the neighborhood of x (see, e.g., [106, 145]).

We assume that the values of the uncertainty function are defined using a *sensory environment*, i.e., a pair $(L, \| \cdot \|_U)$, where L is a set of formulas, called the *sensory formulas*, and $\| \cdot \|_U : L \longrightarrow P(U)$ is the *sensory semantics*. We assume that for any sensory formula α and any object $x \in U$ the information if $x \in \|\alpha\|_U$ holds is available. The set $\{\alpha : x \in \|\alpha\|_U\}$ is called the *signature of x* in AS and is denoted by $Inf_{AS}(x)$. For any $x \in U$ the *set $\mathcal{N}_{AS}(x)$ of neighborhoods of x in AS* is defined by $\mathcal{N}_{AS}(x) = \{\|\alpha\|_U : x \in \|\alpha\|_U\}$ and from this set the neighborhood $I(x)$ is constructed. For example, $I(x)$ is defined by selecting an element from the set $\{\|\alpha\|_U : x \in \|\alpha\|_U\}$ or by $I(x) = \bigcap \mathcal{N}_{AS}(x)$. Observe that any sensory environment $(L, \| \cdot \|_U)$ can be treated as a parameter of I from the vector $\#$ (see Definition 2.12).

Let us consider two examples.

Example 2.13. Any decision table $DT = (U, A, d)$ [106] defines an approximation space $AS_{DT} = (U, I_A, \nu_\$)$, where, as we will see, $I_A(x) = \{y \in U : a(y) = a(x)$

for all $a \in A\}$. Any sensory formula is a descriptor (selector), i.e., a formula of the form $a = v$ where $a \in A$ and $v \in V_a$ with the standard semantics $\|a = v\|_U = \{x \in U : a(x) = v\}$. Then, for any $x \in U$ its signature $Inf_{AS_{DT}}(x)$ is equal to $\{a = a(x) : a \in A\}$ and the neighborhood $I_A(x)$ is equal to $\bigcap \mathcal{N}_{AS_{DT}}(x)$.

Example 2.14. Another example can be obtained assuming that for any $a \in A$ there is given a tolerance relation $\tau_a \subseteq V_a \times V_a$ (see, e.g., [145]). Let $\tau = \{\tau_a\}_{a \in A}$. Then, one can consider a tolerance decision table $DT_\tau = (U, A, d, \tau)$ with tolerance descriptors $a =_{\tau_a} v$ and their semantics $\|a =_{\tau_a} v\|_U = \{x \in U : v\tau_a a(x)\}$. Any such tolerance decision table $DT_\tau = (U, A, d, \tau)$ defines the approximation space $AS_{DT_\tau} = (U, I_A, \nu_{\$})$ with the signature $Inf_{AS_{DT_\tau}}(x) = \{a =_{\tau_a} a(x) : a \in A\}$ and the neighborhood $I_A(x) = \bigcap \mathcal{N}_{AS_{DT_\tau}}(x)$ for any $x \in U$.

The fusion of $\mathcal{N}_{AS_{DT_\tau}}(x)$ for computing the neighborhood of x can have many different forms; the intersection is only an example. One can also consider some more general uncertainty functions, e.g., with values in $P^2(U) = P(P(U))$ [161]. For example, to compute the value of $I(x)$ first some subfamilies of $\mathcal{N}_{AS}(x)$ can be selected and next the family consisting of intersection of each such a subfamily is taken as the value of $I(x)$.

Note, that any sensory environment $(L, \|\cdot\|_U)$ defines an information system with the universe U of objects. Any row of such an information system for an object x consists of information if $x \in \|\alpha\|_U$ holds, for any sensory formula α. Let us also observe that in our examples we have used a simple sensory language defined by descriptors of the form $a = v$. One can consider a more general approach by taking, instead of the simple structure $(V_a, =)$, some other relational structures R_a with the carrier V_a for $a \in A$ and a signature τ. Then any formula (with one free variable) from a sensory language with the signature τ that is interpreted in R_a defines a subset $V \subseteq V_a$ and induces on the universe of objects a neighborhood consisting of all objects having values of the attribute a in the set V.

Example 2.15. Let us define a language L_{IS} used for elementary granule description, where $IS = (U, A)$ is an information system. The syntax of L_{IS} is defined recursively by

1. $(a \text{ in } V) \in L_{IS}$, for any $a \in A$ and $V \subseteq V_a$.
2. If $\alpha \in L_{IS}$ then $\neg \alpha \in L_{IS}$.
3. If $\alpha, \beta \in L_{IS}$ then $\alpha \wedge \beta \in L_{IS}$.
4. If $\alpha, \beta \in L_{IS}$ then $\alpha \vee \beta \in L_{IS}$.

The semantics of formulas from L_{IS} with respect to an information system IS is defined recursively by

1. $\|a \text{ in } V\|_{IS} = \{x \in U : a(x) \in V\}$.
2. $\|\neg \alpha\|_{IS} = U - \|\alpha\|_{IS}$.
3. $\|\alpha \wedge \beta\|_{IS} = \|\alpha\|_{IS} \cap \|\beta\|_{IS}$.
4. $\|\alpha \vee \beta\|_{IS} = \|\alpha\|_{IS} \cup \|\beta\|_{IS}$.

A typical method used by the classical rough set approach [106] for constructive definition of the uncertainty function is the following: for any object $x \in U$ there is given information $Inf_A(x)$ (information vector, attribute value vector of x) which can be interpreted as a conjunction $EF_B(x)$ of selectors $a = a(x)$ for $a \in A$ and the set $I_\#(x)$ is equal to $\|EF_B(x)\|_{IS} = \|\bigwedge_{a \in A} a = a(x)\|_{IS}$. One can consider a more general case taking as possible values of $I_\#(x)$ any set $\|\alpha\|_{IS}$ containing x. Next from the family of such sets the resulting neighborhood $I_\#(x)$ can be selected or constructed. One can also use another approach by considering more general approximation spaces in which $I_\#(x)$ is a family of subsets of U [20], [81].

2.3.2   Rough Inclusion Function

One can consider general constraints which the rough inclusion functions should satisfy. Searching for such constraints initiated investigations resulting in creation and development of rough mereology (see, e.g., [118, 122] and the bibliography in [118]). In this subsection, we present only some examples of rough inclusion functions.

The rough inclusion function $\nu_\$: P(U) \times P(U) \to [0,1]$ defines the degree of inclusion of X in Y, where $X, Y \subseteq U$.

In the simplest case the standard rough inclusion function can be defined by (see, e.g., [106, 145]):

$$\nu_{SRI}(X,Y) = \begin{cases} \frac{card(X \cap Y)}{card(X)} & \text{if } X \neq \emptyset \\ 1 & \text{if } X = \emptyset. \end{cases} \tag{2.1}$$

Some illustrative example is given in Table 2.3.

Table 2.3. Illustration of Standard Rough Inclusion Function

X	Y	$\nu_{SRI}(X,Y)$
$\{x_1, x_3, x_7, x_8\}$	$\{x_2, x_4, x_5, x_6, x_9\}$	0
$\{x_1, x_3, x_7, x_8\}$	$\{x_1, x_2, x_4, x_5, x_6, x_9\}$	0.25
$\{x_1, x_3, x_7, x_8\}$	$\{x_1, x_2, x_3, x_7, x_8\}$	1

This measure is widely used by the data mining and rough set communities. It is worth mentioning that Jan Łukasiewicz [79] was the first one who used this idea to estimate the probability of implications. However, rough inclusion can have a much more general form than inclusion of sets to a degree (see, e.g., [118, 122, 161]).

Another example of rough inclusion function ν_t can be defined using the standard rough inclusion and a threshold $t \in (0, 0.5)$ using the following formula:

$$\nu_t(X,Y) = \begin{cases} 1 & \text{if } \nu_{SRI}(X,Y) \geq 1 - t \\ \frac{\nu_{SRI}(X,Y) - t}{1 - 2t} & \text{if } t \leq \nu_{SRI}(X,Y) < 1 - t \\ 0 & \text{if } \nu_{SRI}(X,Y) \leq t \end{cases} \tag{2.2}$$

The rough inclusion function ν_t is used in the variable precision rough set approach [229].

Another example of rough inclusion is used for function approximation [161] and relation approximation [185].

Then the inclusion function ν^* for subsets $X, Y \subseteq U \times U$, where $X, Y \subseteq \mathcal{R}$ and $\mathcal{R}$ is the set of reals, is defined by

$$\nu^* (X, Y) = \begin{cases} \frac{card(\pi_1(X \cap Y))}{card(\pi_1(X))} & \text{if } \pi_1(X) \neq \emptyset \\ 1 & \text{if } \pi_1(X) = \emptyset \end{cases} \tag{2.3}$$

where π_1 is the projection operation on the first coordinate. Assume now, that X is a cube and Y is the graph $G(f)$ of the function $f : \mathcal{R} \longrightarrow \mathcal{R}$. Then, e.g., X is in the lower approximation of f if the projection on the first coordinate of the intersection $X \cap G(f)$ is equal to the projection of X on the first coordinate. This means that the part of the graph $G(f)$ is "well" included in the box X, i.e., for all arguments that belong to the box projection on the first coordinate the value of f is included in the box X projection on the second coordinate.

Usually, there are several parameters that are tuned in searching for a relevant rough inclusion function. Such parameters are listed in the vector \$. An example of such parameters is the threshold mentioned for the rough inclusion function used in the variable precision rough set model. We would like to mention some other important parameters. Among them are pairs $(L^*, \| \cdot \|_U^*)$ where L^* is an extension of L and $\| \cdot \|_U^*$ is an extension of $\| \cdot \|_U$, where $(L, \| \cdot \|_U)$ is a sensory environment. For example, if L consists of sensory formulas $a = v$ for $a \in A$ and $v \in V_a$ then one can take as L^* the set of descriptor conjunctions. For rule based classifiers we search in such a set of formulas for patterns relevant for decision classes.

2.3.3   Lower and Upper Approximations

The lower and the upper approximations of subsets of U are defined as follows.

Definition 2.16. *For any approximation space* $AS_{\#,\$} = (U, I_\#, \nu_\$)$ *and any subset* $X \subseteq U$, *the lower and upper approximations are defined by*

$$LOW \left(AS_{\#,\$}, X \right) = \{ x \in U : \nu_\$ \left(I_\# \left(x \right), X \right) = 1 \},$$

$$UPP \left(AS_{\#,\$}, X \right) = \{ x \in U : \nu_\$ \left(I_\# \left(x \right), X \right) > 0 \}.$$

The lower approximation of a set X with respect to the approximation space $AS_{\#,\$}$ is the set of all objects, which can be classified with certainty as objects of X with respect to $AS_{\#,\$}$. The upper approximation of a set X with respect to the approximation space $AS_{\#,\$}$ is the set of all objects which can be possibly classified as objects of X with respect to $AS_{\#,\$}$.

The difference between the upper and lower approximation of a given set is called its boundary region:

$$BN \left(AS_{\#,\$}, X \right) = UPP \left(AS_{\#,\$}, X \right) - LOW \left(AS_{\#,\$}, X \right).$$

Rough set theory expresses vagueness by employing a boundary region of a set. If the boundary region of a set is empty it means that the set is crisp, otherwise the set is rough (inexact). A nonempty boundary region of a set indicates that our knowledge about the set is not sufficient to define the set precisely. One can recognize that rough set theory is, in a sense, a formalization of the idea presented by a German mathematician Gotlob Frege (1848–1925) [37].

Several known approaches to concept approximations can be covered using the discussed here approximation spaces, e.g., the approach given in [106], approximations based on the variable precision rough set model [229] or tolerance (similarity) rough set approximations (see, e.g., [145] and references in [145]).

Rough sets can approximately describe sets of patients, events, outcomes, keywords, etc. that may be otherwise difficult to circumscribe.

Example 2.17. Let U be a set of patients and we consider two attributes a_2 (age of disease diagnosis) and a_3 (disease duration) (see Example 2.10 and Figure 2.2).
Let
$$V_{a_2} = \{preschool, early\_school, adolescence\}$$
and
$$V_{a_3} = \{short, medium, long\}.$$
In this case we obtain nine granules corresponding to conjunctions of descriptors e.g.

$$(a_2, preschool) \wedge (a_3, medium), (a_2, adolescence) \wedge (a_3, short), \ldots$$

For a set X of patients the lower and the upper approximation is also depicted on Figure 2.2.

Example 2.18. We consider parameterized approximation spaces in information retrieval problem [38, 39]. At first, in order to determine an approximation space, we choose the universe U as the set of all keywords. Let DOC be a set of documents, which are described by keywords. Let $key : DOC \longrightarrow P(U)$ be a function mapping documents into sets of keywords. Denote by $c(x_i, x_j)$, where $c : U \times U \longrightarrow \{0, 1, 2, \ldots\}$, the frequency of co-occurrence between two keywords x_i and x_j i.e.

$$c(x_i, x_j) = card(\{doc \in DOC : \{x_i, x_j\} \subseteq key(doc)\}).$$

We define the uncertainty function I_θ depending on a threshold $\theta \in \{0, 1, \ldots\}$ as follows:
$$I_\theta(x_i) = \{x_j \in U : c(x_i, x_j) \geq \theta\} \cup \{x_i\}.$$

One can consider the standard rough inclusion function.

A query is defined as a set of keywords. Different strategies of information retrieval based on the lower and the upper approximations of queries and documents are investigated in [38, 39].

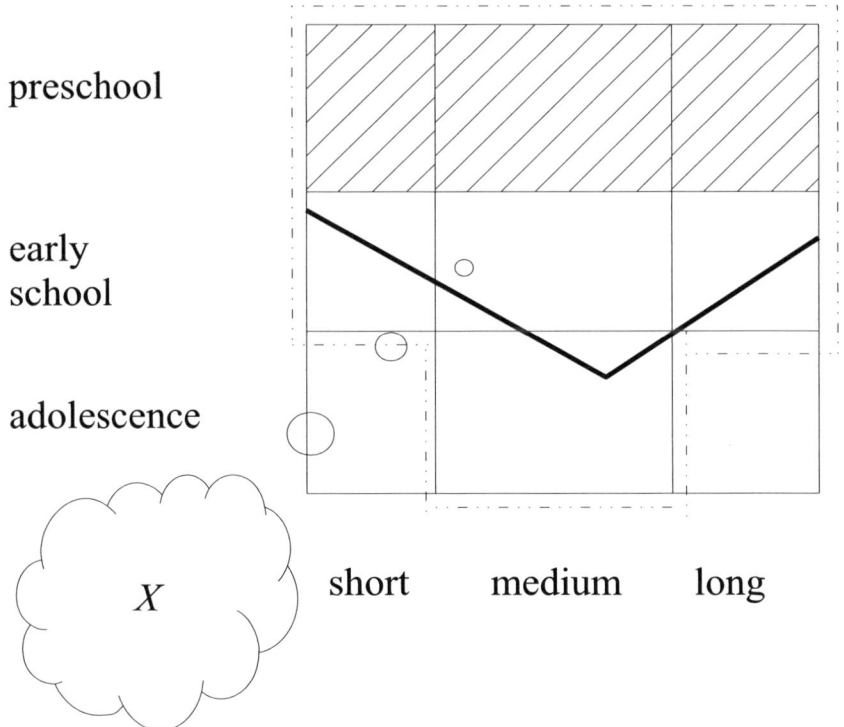

preschool

early
school

adolescence

X short   medium   long

Fig. 2.2. Approximations in the Standard Rough Set Model

The classification methods for concept approximation developed in machine learning and pattern recognition make it possible to decide if a given object belongs to the approximated concept or not [47]. The classification methods yield the decisions using only partial information about approximated concepts. This fact is reflected in the rough set approach by assumption that concept approximations should be defined using only partial information about approximation spaces. To decide if a given object belongs to the (lower or upper) approximation of a given concept the rough inclusion function values are needed. In the next section, we show how such values, so needed in classification making, are estimated on the basis of available partial information about approximation spaces.

2.3.4 Properties of Approximations

A rough set can be also characterized numerically by the coefficient called the accuracy of approximation.

Definition 2.19. *The accuracy of approximation is equal to the number*

$$\alpha\left(AS_{\#,\$},X\right) = \frac{card\left(LOW\left(AS_{\#,\$},X\right)\right)}{card\left(UPP\left(AS_{\#,\$},X\right)\right)}.$$

If $\alpha\left(AS_{\#,\$}, X\right) = 1$, then X is *crisp* with respect to $AS_{\#,\$}$ (X is *precise* with respect to $AS_{\#,\$}$), and otherwise, if $\alpha\left(AS_{\#,\$}, X\right) < 1$, then X is *rough* with respect to $AS_{\#,\$}$ (X is *vague* with respect to $AS_{\#,\$}$).

We recall the notions of the positive region and the quality of approximation of classification in the case of generalized approximation spaces.

Definition 2.20. *Let $AS_{\#,\$} = (U, I_{\#}, \nu_{\$})$ be an approximation space. Let $\{X_1, \ldots, X_r\}$ be a classification of objects (i.e. $X_1, \ldots, X_r \subseteq U$, $\bigcup_{i=1}^{r} X_i = U$ and $X_i \cap X_j = \emptyset$ for $i \neq j$, where $i, j = 1, \ldots, r$).*

1. *The positive region of the classification $\{X_1, \ldots, X_r\}$ with respect to the approximation space $AS_{\#,\$}$ is defined by*

$$POS\left(AS_{\#,\$}, \{X_1, \ldots, X_r\}\right) = \bigcup_{i=1}^{r} LOW\left(AS_{\#,\$}, X_i\right).$$

2. *The quality of approximation of the classification $\{X_1, \ldots, X_r\}$ in the approximation space $AS_{\#,\$}$ is defined by*

$$\gamma\left(AS_{\#,\$}, \{X_1, \ldots, X_r\}\right) = \frac{card\left(POS\left(AS_{\#,\$}, \{X_1, \ldots, X_r\}\right)\right)}{card\left(U\right)}.$$

The positive region for three decision classes is depicted on Figure 2.3.

The quality of approximation of the classification coefficient expresses the ratio of the number of all $AS_{\#,\$}$-correctly classified objects to the number of all objects in the data table. In other words

$$\frac{\text{Number of objects in lower approximations}}{\text{Total number of objects}}.$$

Now, we list properties of approximations in generalized approximation spaces. Next, we present definitions and give an idea of algorithms for checking rough definability, internal undefinability etc.

Let $AS = (U, I, \nu)$ be an approximation space. For two sets $X, Y \subseteq U$ the equality with respect to the rough inclusion ν is defined in the following way: $X =_{\nu} Y$ if and only if $\nu\left(X, Y\right) = 1 = \nu\left(Y, X\right)$.

Proposition 2.21. *Assuming that for every $x \in U$ we have $x \in I\left(x\right)$ and that ν_{SRI} is the standard rough inclusion one can show the following properties of approximations:*

1. *$\nu_{SRI}\left(LOW\left(AS, X\right), X\right) = 1$ and $\nu_{SRI}\left(X, UPP\left(AS, X\right)\right) = 1$.*
2. *$LOW\left(AS, \emptyset\right) =_{\nu_{SRI}} UPP\left(AS, \emptyset\right) =_{\nu_{SRI}} \emptyset$.*
3. *$LOW\left(AS, U\right) =_{\nu_{SRI}} UPP\left(AS, U\right) =_{\nu_{SRI}} U$.*
4. *$UPP\left(AS, X \cup Y\right) =_{\nu_{SRI}} UPP\left(AS, X\right) \cup UPP\left(AS, Y\right)$.*
5. *$\nu_{SRI}\left(UPP\left(AS, X \cap Y\right), UPP\left(AS, X\right) \cap UPP\left(AS, Y\right)\right) = 1$.*
6. *$LOW\left(AS, X \cap Y\right) =_{\nu_{SRI}} LOW\left(AS, X\right) \cap LOW\left(AS, Y\right)$.*
7. *$\nu_{SRI}\left(LOW\left(AS, X\right) \cup LOW\left(AS, Y\right), LOW\left(AS, X \cup Y\right)\right) = 1$.*
8. *$\nu_{SRI}\left(X, Y\right) = 1$ implies $\nu_{SRI}\left(LOW\left(AS, X\right), LOW\left(AS, Y\right)\right) = 1$.*

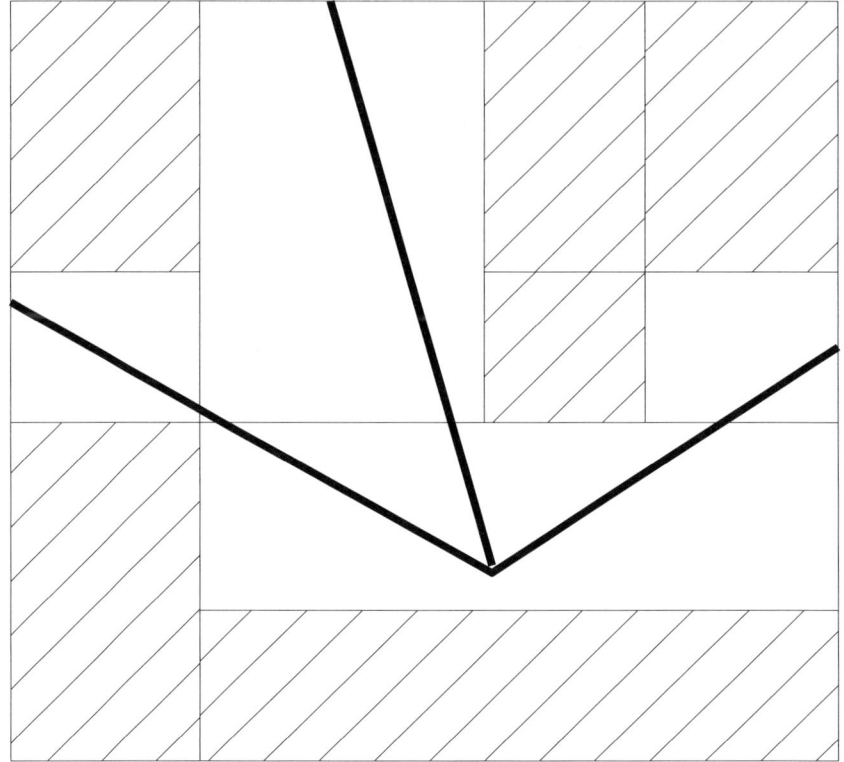

Fig. 2.3. Positive Region

9. $\nu_{SRI}(X,Y) = 1$ implies $\nu_{SRI}(UPP(AS,X),UPP(AS,Y)) = 1$.
10. $LOW(AS, U - X) =_{\nu_{SRI}} U - UPP(AS,X)$.
11. $UPP(AS, U - X) =_{\nu_{SRI}} U - LOW(AS,X)$.
12. $\nu_{SRI}(LOW(AS, LOW(AS,X)), LOW(AS,X)) = 1$.
13. $\nu_{SRI}(LOW(AS,X), UPP(AS, LOW(AS,X))) = 1$.
14. $\nu_{SRI}(LOW(AS, UPP(AS,X)), UPP(AS,X)) = 1$.
15. $\nu_{SRI}(UPP(AS,X), UPP(AS, UPP(AS,X))) = 1$.

By analogy with the standard rough set theory, we define the following four types of sets:

1. X is *roughly AS-definable* if and only if $LOW(AS,X) \neq_{\nu_{SRI}} \emptyset$ and $UPP(AS,X) \neq_{\nu_{SRI}} U$.
2. X is *internally AS-undefinable* if and only if $LOW(AS,X) =_{\nu_{SRI}} \emptyset$ and $UPP(AS,X) \neq_{\nu_{SRI}} U$.
3. X is *externally AS-undefinable* if and only if $LOW(AS,X) \neq_{\nu_{SRI}} \emptyset$ and $UPP(AS,X) =_{\nu_{SRI}} U$.
4. X is *totally AS-undefinable* if and only if $LOW(AS,X) =_{\nu_{SRI}} \emptyset$ and $UPP(AS,X) =_{\nu_{SRI}} U$.

The intuitive meaning of this classification is the following.

If X is roughly AS-definable, then with the help of AS we are able to decide for some elements of U whether they belong to X or $U - X$.

If X is internally AS-undefinable, then we are able to decide whether some elements of U belong to $U - X$, but we are unable to decide for any element of U, whether it belongs to X, using AS.

If X is externally AS-undefinable, then we are able to decide for some elements of U whether they belong to X, but we are unable to decide, for any element of U whether it belongs to $U - X$, using AS.

If X is totally AS-undefinable, then we are unable to decide for any element of U whether it belongs to X or $U - X$, using AS.

The algorithms for checking corresponding properties of sets have $O\left(n^2\right)$ time complexity, where $n = card\left(U\right)$. Let us also note that using two properties of approximations:

$$LOW\left(AS, U - X\right) =_{\nu_{SRI}} U - UPP\left(AS, X\right),$$

$$UPP\left(AS, U - X\right) =_{\nu_{SRI}} U - LOW\left(AS, X\right)$$

one can obtain internal AS-undefinability of X if and only if $U - X$ is externally AS-undefinable. Having that property, we can utilize an algorithm that check internal undefinability of X to examine if $U - X$ is externally undefinable.

2.4   Rough Relations

One can distinguish several directions in research on relation approximations. Below we list some examples of them. In [105], [177] properties of the rough relations are presented. The relationships of rough relations and modal logics have been investigated by many authors (see e.g. [207], [138]). We refer to [138], where the upper approximation of the input-output relation $R(P)$ of a given program P with respect to indiscernibility relation IND is treated as the composition $IND \circ R\left(P\right) \circ IND$ and where a special symbol for the lower approximation of $R\left(P\right)$ is introduced. Properties of relation approximations in generalized approximation spaces are presented in [141], [186]. The relationships of rough sets with algebras of relations are investigated for example in [101], [29]. Relationships between rough relations and a problem of objects ranking are presented for example in [43], where it is shown that the classical rough set approximations based on indiscernibility relation do not take into account the ordinal properties of the considered criteria. This drawback is removed by considering rough approximations of the preference relations by graded dominance relations [43] and generally, dominance based rough set approach [164]. In [98] some properties of rough relations found in the literature were proved.

In this section we discuss approximations of relations with respect to different rough inclusions. For simplicity of the presentation we consider only binary relations.

Let $AS = (U, I, \nu)$ be an approximation space, where $U \subseteq U_1 \times U_2$ and U, U_1, U_2 are non-empty sets.

By $\pi_i(R)$ we denote the projection of the relation $R \subseteq U$ onto the $i-th$ axis i.e. for example for $i = 1$

$$\pi_1(R) = \{x_1 \in U_1 : \exists_{x_2 \in U_2}(x_1, x_2) \in R\}.$$

Definition 2.22. *For any relations $S, R \subseteq U$ the rough inclusion functions ν_{π_1} and ν_{π_2} based on the cardinality of the projections are defined as follows:*

$$\nu_{\pi_i}(S, R) = \begin{cases} \frac{card(\pi_i(S \cap R))}{card(\pi_i(S))} & \text{if } S \neq \emptyset \\ 1 & \text{if } S = \emptyset \end{cases},$$

where $i = 1, 2$.

We describe the intuitive meaning of the approximations in approximation spaces $AS_\$ = (U, I, \nu_\$)$, where $\$ \in \{SRI, \pi_1, \pi_2\}$. The standard lower approximation $LOW(AS_{SRI}, R)$ of a relation $R \subseteq U$ has the following property: any objects $(x_1, x_2) \in U$ are connected by the lower approximation of R if and only if any objects (y_1, y_2) from $I((x_1, x_2))$ are in the relation R. One can obtain some less restrictive definitions of the lower approximation using the rough inclusions ν_{π_1} and ν_{π_2}. The pair (x_1, x_2) is in the lower approximation $LOW(AS_{\pi_1}, R)$ if and only if for every y_1 there is y_2 such that the pair (y_1, y_2) is from $I((x_1, x_2)) \cap R$. One can obtain similar interpretation for ν_{π_2}. The upper approximation with respect to all introduced rough inclusions is exactly the same, namely, the pair $(x_1, x_2) \in U$ is in the upper approximation $UPP(AS_\$, R)$, where $\$ \in \{SRI, \pi_1, \pi_2\}$ if and only if there is a pair (y_1, y_2) from $I((x_1, x_2)) \cap R$.

Proposition 2.23. *For the lower and the upper approximations the following conditions are satisfied:*

1. $LOW(AS_{SRI}, R) \subseteq R$.
2. $LOW(AS_{SRI}, R) \subseteq LOW(AS_{\pi_1}, R)$.
3. $LOW(AS_{SRI}, R) \subseteq LOW(AS_{\pi_2}, R)$.
4. $R \subseteq UPP(AS_{SRI}, R) = UPP(AS_{\pi_1}, R) = UPP(AS_{\pi_2}, R)$.

Example 2.24. We give some example which illustrates that the inclusions from the last proposition can not to be replaced by equalities. Let us also observe that the universe U need not be equal to the Cartesian product of two sets. Let the

Table 2.4. Uncertainty Function and Rough Inclusions

U	I	ν_{SRI}	ν_{π_1}	ν_{π_2}
$(1, 2)$	$\{(1, 2), (1, 3)\}$	0.5	1	0.5
$(2, 1)$	$\{(2, 1), (2, 3), (3, 1)\}$	0.67	0.5	1
$(2, 3)$	$\{(2, 1), (2, 3), (3, 1)\}$	0.67	0.5	1
$(3, 2)$	$\{(3, 2)\}$	1	1	1
$(1, 3)$	$\{(1, 2), (1, 3)\}$	0.5	1	0.5
$(3, 1)$	$\{(2, 1), (2, 3), (3, 1)\}$	0.67	0.5	1

Table 2.5. Approximations

$LOW\,(AS_{SRI}, R)$	$\{(3,2)\}$
$LOW\,(AS_{\pi_1}, R)$	$\{(1,2),(3,2),(1,3)\}$
$LOW\,(AS_{\pi_2}, R)$	$\{(2,1),(2,3),(3,2),(3,1)\}$
$UPP\,(AS_\$, R)$	U

universe $U = \{(1,2),(2,1),(2,3),(3,2),(1,3),(3,1)\}$ and the binary relation $R = \{(1,2),(2,1),(2,3),(3,2)\}$.

The definition of an uncertainty function $I : U \rightarrow P(U)$ and the rough inclusions are described in Table 2.4.

The lower and the upper approximations of R in the approximation spaces $AS_\$ = (U, I, \nu_\$)$, where $\$ \in \{SRI, \pi_1, \pi_2\}$ are described in Table 2.5.

Proposition 2.25. *The time complexity of algorithms for computing approximations of relations is equal to* $O\left((card\,(U))^2\right)$.

2.5   Function Approximation

In this section, we are looking for the high quality (in the rough set framework) of function approximation from available incomplete data. Our approach can be treated as a kind of rough clustering of functional data.

Let us consider an example of function approximation. We assume that a partial information is only available about a function, this means that, some points from the graph of the function are known. We would like to present a more formal description of function approximation. The application of this concept for definition of rough-integral over partially specified functions is given in [60].

First, let us introduce some notation. Let us assume U^∞ is the universe of objects and we assume that μ is a measure on a σ-field of subsets of U^∞. By $U \subseteq U^\infty$ we denote a finite sample (training set) of objects from U^∞. We assume that $\mu(U^\infty) < \infty$. By R_+ we denote the set of non-negative reals and by μ_0 a measure on a σ-field of subsets of R_+. A function $f : U \longrightarrow R_+$ will be called a sample of a function $f^* : U^\infty \rightarrow R_+$ if f^* is an extension of f. For any $Z \subseteq U^\infty \times R_+$ by $\pi_1(Z)$ and $\pi_2(Z)$ we denote the set $\{x \in U^\infty : \exists y \in R_+ \ (x,y) \in Z\}$ and $\{y \in R_+ : \exists x \in U^\infty \ (x,y) \in Z\}$, respectively.

If $\mathcal{C}$ is a family of neighborhoods, i.e., non-empty subsets of $U^\infty \times R_+$ (measurable relative to the product measure $\mu \times \mu_0$) then the lower approximation of f relative to approximation space $AS_\mathcal{C}$ (see Figure 2.4 where neighborhoods marked by solid lines belong to the lower approximation and with dashed lines - to the upper approximation) is defined by

$$LOW(AS_\mathcal{C}, f) = \bigcup \{c \in \mathcal{C} : \ f(\pi_1(c) \cap U) \subseteq \pi_2(c)\}. \tag{2.4}$$

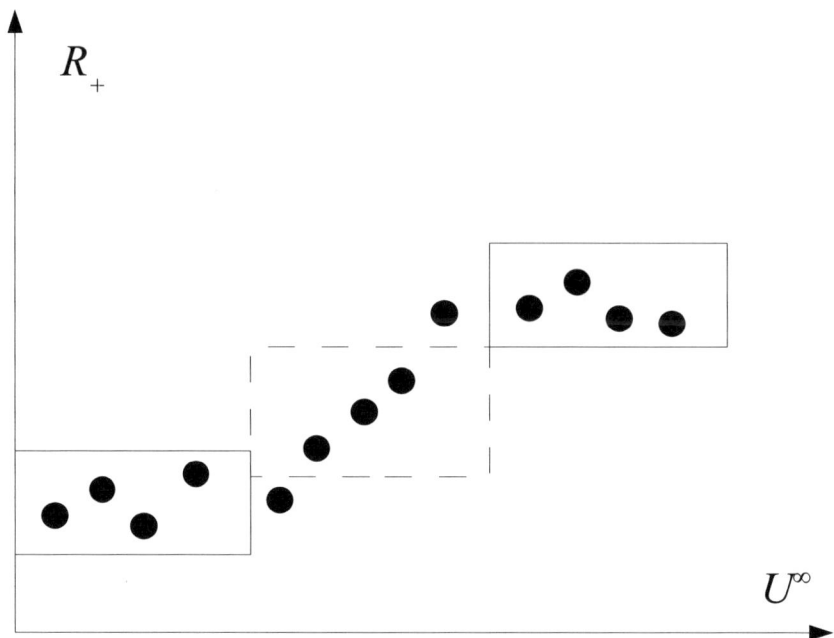

Fig. 2.4. Function Approximation

Observe that this definition is different from the standard definition of the lower approximation [106, 108]. We can easily see that if we apply the definition of relation approximation to f (it is a special case of relation) then the lower approximation is almost always empty. The new definition is making it possible express better the fact that the graph of f is "well" matching a given neighborhood [158]. For expressing this a classical set theoretical inclusion of neighborhood into the graph of f is not satisfactory.

One can define the upper approximation of f relative to $AS_\mathcal{C}$ by

$$UPP(AS_\mathcal{C}, f) = \bigcup \{c \in \mathcal{C} : f(\pi_1(c) \cap U) \cap \pi_2(c) \neq \emptyset\}. \qquad (2.5)$$

In applications, neighborhoods are defined constructively by semantics of some formulas. Let us assume that $\mathcal{F}$ is a given set of formulas and for any formula $\alpha \in \mathcal{F}$ there are defined two semantics: $\|\alpha\|_U \subseteq U \times R_+$ and $\|\alpha\|_{U^\infty} \subseteq U^\infty \times R_+$, i.e., semantics on the sample U and on the whole universe U^∞. We obtain two families of neighborhoods $\mathcal{F}_U = \{\|\alpha\|_U \subseteq U \times R_+ : \alpha \in \mathcal{F}\}$ and $\mathcal{F}_{U^\infty} = \{\|\alpha\|_{U^\infty \times R_+} : \alpha \in \mathcal{F}\}$. To this end, we consider (measurable) neighborhoods of the form $Z \times I$ where $Z \subseteq U^\infty$ and I is an interval of reals.

We know that $\|\alpha\|_U = \|\alpha\|_{U^\infty} \cap (U \times R_+)$ but having the sample we do not have information about the other objects from $U^\infty \setminus U$. Hence, for defining the lower approximation of f over U^∞ on the basis of the lower approximation over U some estimation methods should be used.

Example 2.26. We present an illustrative example of a function $f : U \to R_+$ approximation where $U = \{1, 2, 4, 5, 7, 8\}$. Let $f(1) = 3$, $f(2) = 2$, $f(4) = 2$, $f(5) = 5$, $f(7) = 5$, $f(8) = 2$. We consider three indiscernibility classes $C_1 = [0, 3] \times [1.5, 4]$, $C_2 = [3, 6] \times [1.7, 4.5]$ and $C_3 = [6, 9] \times [3, 4]$. We compute projections of indiscernibility classes: $\pi_1(C_1) = [0, 3]$, $\pi_2(C_1) = [1.5, 4]$, $\pi_1(C_2) = [3, 6]$, $\pi_2(C_2) = [1.7, 4.5]$, $\pi_1(C_3) = [6, 9]$ and $\pi_2(C_3) = [3, 4]$. Hence we obtain $f(\pi_1(C_1) \cap U) = f(\{1, 2\}) = \{2, 3\} \subseteq \pi_2(C_1)$, $f(\pi_1(C_2) \cap U) = f(\{4, 5\}) = \{2, 5\} \nsubseteq \pi_2(C_2)$ but $f(\pi_1(C_2) \cap U) \cap \pi_2(C_2) = \{2, 5\} \cap [1.7, 4.5] \neq \emptyset$, $f(\pi_1(C_3) \cap U) = \emptyset$. We obtain the lower approximation $LOW(AS_C, f) = C_1$ and the upper approximation $UPP(AS_C, f) = C_1 \cup C_2$.

On can extend the discussed approach to function approximation for the case when instead of the partial graph of a function it is given a more general information consisting of many possible values for a given $x \in U$ due to repetitive measurements influenced by noise.

2.6   Quality of Approximation Space

A key task in granular computing is the information granulation process that leads to the formation of information aggregates (patterns) from a set of available objects. A methodological and algorithmic issue is the formation of transparent (understandable) information granules inasmuch as they should provide a clear and understandable description of patterns present in sample objects [3, 113]. Such a fundamental property can be formalized by a set of constraints that must be satisfied during the information granulation process. Usefulness of these constraints is measured by quality of an approximation space:

$$Quality_1 : Set\_AS \times P(U) \to [0, 1],$$

where U is a non-empty set of objects and Set_AS is a set of possible approximation spaces with the universe U.

Example 2.27. If the upper approximation $UPP(AS, X)) \neq \emptyset$ for $AS \in Set\_AS$ and $X \subseteq U$ then

$$Quality_1(AS, X) = \nu_{SRI}(UPP(AS, X), LOW(AS, X)) =$$

$$\frac{card(UPP(AS, X) \cap LOW(AS, X))}{card(UPP(AS, X))} = \frac{card(LOW(AS, X))}{card(UPP(AS, X))}.$$

The value $1 - Quality_1(AS, X)$ expresses the degree of completeness of our knowledge about X, given the approximation space AS. This value is also called the roughness of the set X with respect to AS. If roughness of the set X is 0 then X is crisp with respect to AS, and if $Quality_1(AS, X) < 1$ then X is rough (i.e., X is vague with respect to AS).

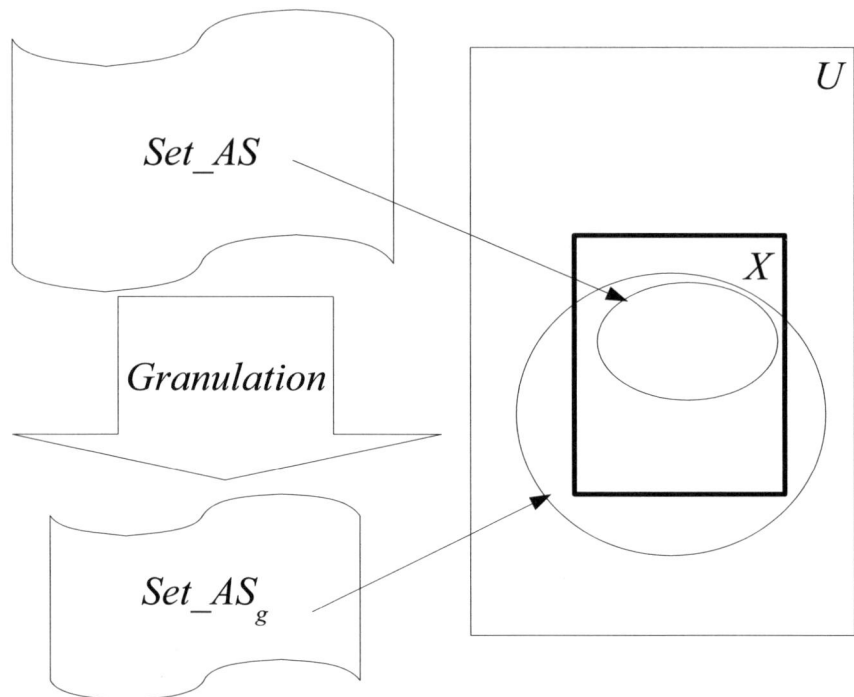

Fig. 2.5. Granulation of Parameterized Approximation Spaces

Example 2.28. In applications, we usually use another quality measure analogous to the minimal length principle [128, 202] where also the description length of approximation is included. Let us denote by $description(AS, X)$ the description length of approximation of X in AS. The description length may be measured, *e.g.*, by the sum of description lengths of algorithms testing membership for neighborhoods used in construction of the lower approximation, the upper approximation, and the boundary region of the set X. Then the quality $Quality_2(AS, X)$ can be defined by

$$Quality_2(AS, X) = g(Quality_1(AS, X), description(AS, X)),$$

where g is a relevant function used for fusion of values $Quality_1(AS, X)$ and $description(AS, X)$. This function g can reflect weights given by experts relative to both criteria.

One can consider different optimization problems relative to a given class Set_AS of approximation spaces. For example, for a given $X \subseteq U$ and a threshold $t \in [0, 1]$, one can search for an approximation space AS satisfying the constraint $Quality_2(AS, X) \geq t$.

Another example can be related to searching for an approximation space satisfying additionally the constraint $Cost(AS) < c$ where $Cost(AS)$ denotes the cost of approximation space AS (*e.g.*, measured by the number of attributes used to define neighborhoods in AS) and c is a given threshold. In the following example, we consider also costs of searching for relevant approximation spaces in a given family defined by a parameterized approximation space (see Figure 2.5). Any parameterized approximation space $AS_{\#,\$} = (U, I_\#, \nu_\$)$ is a family of approximation spaces. The cost of searching in such a family for a relevant approximation space for a given concept X approximation can be treated as a factor of the quality measure of approximation of X in $AS_{\#,\$} = (U, I_\#, \nu_\$)$. Hence, such a quality measure of approximation of X in $AS_{\#,\$}$ can be defined by

$$Quality_3(AS_{\#,\$}, X) = h(Quality_2(AS, X), Cost\_Search(AS_{\#,\$}, X)),$$

where AS is the result of searching in $AS_{\#,\$}$, $Cost\_Search(AS_{\#,\$}, X)$ is the cost of searching in $AS_{\#,\$}$ for AS, and h is a fusion function, *e.g.*, assuming that the values of $Quality_2(AS, X)$ and $Cost\_Search(AS_{\#,\$}, X)$ are normalized to interval $[0, 1]$ h could be defined by a linear combination of $Quality_2(AS, X)$ and $Cost\_Search(AS_{\#,\$}, X)$ of the form

$$\lambda Quality_2(AS, X) + (1 - \lambda)Cost\_Search(AS_{\#,\$}, X),$$

where $0 \leq \lambda \leq 1$ is a weight measuring an importance of quality and cost in their fusion.

We assume that the fusion functions g, h in the definitions of quality are monotonic relative to each argument.

Let $AS \in Set\_AS$ be an approximation space relevant for approximation of $X \subseteq U$, *i.e.*, AS is the optimal (or semi-optimal) relative to $Quality_2$. By $Granulation(AS_{\#,\$})$ we denote a new parameterized approximation space obtained by granulation of $AS_{\#,\$}$. For example, $Granulation(AS_{\#,\$})$ can be obtained by reducing the number of attributes or inclusion degrees (*i.e.*, possible values of the inclusion function). Let AS' be an approximation space in $Granulation(AS_{\#,\$})$ obtained as the result of searching for optimal (semi-optimal) approximation space in $Granulation(AS_{\#,\$})$ for approximation of X.

We assume that three conditions are satisfied:

- after granulation of $AS_{\#,\$}$ to $Granulation(AS_{\#,\$})$ the following property holds: the cost

$$Cost\_Search(Granulation(AS_{\#,\$}), X),$$

  is much lower than the cost $Cost\_Search(AS_{\#,\$}, X)$;
- $description(AS', X)$ is much shorter than $description(AS, X)$, *i.e.*, the description length of X in the approximation space AS' is much shorter than the description length of X in the approximation space AS;
- $Quality_1(AS, X)$ and $Quality_1(AS', X)$ are sufficiently close.

The last two conditions should guarantee that the values $Quality_2(AS, X)$ and $Quality_2(AS', X)$ are comparable and this condition together with the first condition about the cost of searching should assure that

$$Quality_3(Granulation(AS_{\#,\$}, X)) \text{ is much better than } Quality_3(AS_{\#,\$}, X).$$

Taking into account that the parameterized approximation spaces are examples of parameterized granules one can generalize the above example of parameterized approximation space granulation to the case of granulation of parameterized granules.

In the process of searching for (sub-)optimal approximation spaces different strategies are used. Let us consider one illustrative example [160]. Let $DT = (U, A, d)$ be a decision system (a given sample of data) where U is a set of objects, A is a set of attributes and d is a decision. We assume that for any object x, there is accessible only partial information equal to the A-signature of x (object signature, for short), *i.e.*, $Inf_A(x) = \{(a, a(x)) : a \in A\}$ and analogously for any concept there is only given a partial information about this concept by a sample of objects, *e.g.*, in the form of decision table. One can use object signatures as new objects in a new relational structure $\mathcal{R}$. In this relational structure $\mathcal{R}$ are also modeled some relations between object signatures, *e.g.*, defined by the similarities of these object signatures. Discovery of relevant relations on object signatures is an important step in the searching process for relevant approximation spaces. In this way, a class of relational structures representing perception of objects and their parts is constructed. In the next step, we select a language $\mathcal{L}$ of formulas expressing properties over the defined relational structures and we search for relevant formulas in $\mathcal{L}$. The semantics of formulas (*e.g.*, with one free variable) from $\mathcal{L}$ are subsets of object signatures. Observe that each object signature defines a neighborhood of objects from a given sample (*e.g.*, decision table DT) and another set on the whole universe of objects being an extension of U. In this way, each formula from $\mathcal{L}$ defines a family of sets of objects over the sample and also another family of sets over the universe of all objects. Such families can be used to define new neighborhoods of a new approximation space, *e.g.*, by taking unions of the described above families. In the searching process for relevant neighborhoods, we use information encoded in the given sample. More relevant neighborhoods are making it possible to define relevant approximation spaces (from the point of view of the optimization criterion). It is worth to mention that often this searching process is even more compound. For example, one can discover several relational structures (not only one, *e.g.*, $\mathcal{R}$ as it was presented before) and formulas over such structures defining different families of neighborhoods from the original approximation space and next fuse them for obtaining one family of neighborhoods or one neighborhood in a new approximation space. This kind of modeling is typical for hierarchical modeling [9], *e.g.*, when we search for a relevant approximation space for objects composed from parts for which some relevant approximation spaces have been already found.

2.7   Learning Approximation Space from Data

In this section we consider problem of learning approximation space from data. The searching for proper approximation space is the crucial and the most difficult task related to decision algorithm synthesis based on approximation spaces.

2.7.1   Discretization and Approximation Spaces

Discretization is considered for real or integer valued attributes. Discretization is based on searching for "cuts" that determine intervals. All values that lie within each interval are then treated as indiscernible. Thus the uncertainty function for an attribute $a \in A$ is defined as shown below.

$y \in I_a(x)$ if and only if values $a(x)$ and $a(y)$ are from the same interval.

Relations obtained on attribute values by using discretization are reflexive, symmetric and transitive.

A simple discretization process consists of the following two steps:

1. Deciding the number of intervals, which is usually done by a user.
2. Determining the width of these intervals, which is usually done by the discretization algorithm.

Example 2.29. The equal width discretization algorithm first finds the minimal and maximal values for each attribute. Then it divides this range of the attribute value into a number of user specified, equal width intervals.

Example 2.30. In equal frequency discretization the algorithm first sorts the values of each attribute in an ascending order, and then divides them into the user specified number of intervals, in such a way that each interval contains the same number of sorted sequential attribute values.

These methods are applied to each attribute independently. They make no use of decision class information.

Several more sophisticated algorithms for automatic discretization (with respect to different optimization criteria) exist, for example one rule discretization [52], Boolean reasoning discretization [95]. For overviews of different discretization methods and discussion of computational complexity of discretization problems see [95]. We outline here only Boolean reasoning discretization. Let $DT = (U, A \cup \{d\})$ be a decision table. For the sake of simplifying the exposition, we will assume that all condition attributes A are numerical. First, for each attribute $a \in A$ we can sort its value set V_a to obtain the following ordering:

$$min_a < \ldots < v_a^i < v_a^{i+1} < \ldots max_a.$$

Next, we place cuts in the middle of $\left[v_a^i, v_a^{i+1}\right]$, except for the situation when all objects that have these values also have equal generalized decision values. Boolean reasoning discretization is based on combining the cuts found by above procedure with Boolean reasoning procedure for discarding all but a small number of cuts such that the discernibility of objects in DT is preserved. The set of solutions to the problem of finding minimal subsets of cuts that preserve the original discernibility of objects in DT with respect to the decision attribute d, are defined by means of the prime implicants of a suitable Boolean function. Often we are interested in employing as few cuts as possible. Then the set covering heuristic is typically used to arrive at a single solution.

Sometimes the best result is obtained by manual discretization. For example, in Chapter 6 discretization of numeric attributes in real life diabetes data was done manually according to medical norms.

2.7.2   Distances and Approximation Spaces

Other approach to searching for an uncertainty function is based on the assumption that there are given some distances on attribute values. Let $a \in A$ be an attribute. We assume that in the set V_a a distance function $\delta_a : V_a \times V_a \to R_+$ (by R_+ we denote the set of non-negative reals) is defined. The distance function δ_a is assumed to satisfy the axioms of a pseudometric, i.e., for any values $v_1, v_2, v_3 \in V_a$:

1. $\delta_a(v_1, v_2) \geq 0$ (non-negativity condition),
2. $\delta_a(v_1, v_1) = 0$ (reflexivity),
3. $\delta_a(v_1, v_2) = \delta_a(v_2, v_1)$ (symmetry),
4. $\delta_a(v_1, v_2) + \delta_a(v_2, v_3) \geq \delta_a(v_1, v_3)$ (triangular inequality).

The distance function δ_a models the relation of similarity between attribute values. The properties of symmetry and triangular inequality are not necessary to model similarity but they are fundamental for the efficiency of many learning methods described in the literature [216]. Relations obtained on attribute values by using distances are at least reflexive. For review of different distances defined on attribute values see [214]. Here we only present some examples of such distances.

Example 2.31. Let $(U, A \cup \{d\})$ be a decision table. One can use the overlap metric, which defines the distance for an attribute as 0 if the values are equal, or 1 if they are different, regardless of which two values they are. The function *overlap* defines the distance between two values v and v' of an attribute $a \in A$ as:

$$overlap_a (v, v') = \begin{cases} 1 \text{ if } v \neq v' \\ 0 \text{ if } v = v' \end{cases}.$$

Example 2.32. The Value Difference Metric (VDM) provides an appropriate distance function for nominal attributes. A version of VDM (without the weighting schemes but with normalization into zero-one interval) defines the distance between two values v and v' of an attribute $a \in A$ as:

$$vdm_a (v, v') = \frac{1}{r(d)} \sum_{i=1}^{r(d)} (\nu_{SRI} (X_v, X_i) - \nu_{SRI} (X_{v'}, X_i))^2,$$

where $r(d)$ is a number of decision classes, ν_{SRI} is the standard rough inclusion, $X_i = \{x \in U : d(x) = i\}$ and $X_v = \{x \in U : a(x) = v\}.$

Using the distance measure VDM, two values are considered to be closer if they have more similar classifications. For example, if an attribute color has three values red, green and blue, and the application is to identify whether or not an object is an apple, then red and green would be considered closer than red and blue because the former two have correlations with decision apple.

If this distance function is used directly on continuous attributes, all values can potentially be unique. Some approaches to the problem of using VDM on continuous attributes are presented in [214].

Example 2.33. One can also use some other distance function for real or integer valued attributes, for example

$$diff_a(v, v') = \frac{|v - v'|}{\max_a - \min_a},$$

where $\max_a$ and $\min_a$ are the maximum and minimum values, respectively, for an attribute $a \in A$. The normalization of numerical attributes with the range of values $\max_a - \min_a$ makes numerical and nominal attributes equally significant.

Example 2.34. One can also use the difference between attribute values defined as follows

$$difference_a(v, v') = \begin{cases} |v - v'| & \text{if} & a \text{ is continuous} \\ 0 & \text{if} & a \text{ is nominal and } v = v' \\ 1 & \text{otherwise} \end{cases} \quad (2.6)$$

One should specify in searching for optimal uncertainty function at least two elements:

- a class of parameterized uncertainty functions,
- an optimization criterion.

Definition 2.35. *Let $\delta_a : V_a \times V_a \longrightarrow [0, \infty)$ be a given distance function on attribute values, where V_a is the set of all values of attribute $a \in A$. One can define the following uncertainty function*

$$y \in I_a^{f_a}(x) \text{ if and only if } \delta_a(a(x), a(y)) \leq f_a(a(x), a(y)),$$

where $f_a : V_a \times V_a \to [0, \infty)$ is a threshold function.

Example 2.36. We present two examples of a threshold function f_a.

1. For every $x, y \in U$ one can define $f_a(a(x), a(y)) = \varepsilon_a$, where $0 \leq \varepsilon_a \leq 1$ is a given real number.
2. One can define for real or integer valued attribute $a \in A$,

$$f_a(a(x), a(y)) = \varepsilon_a^\alpha * a(x) + \varepsilon_a^\beta * a(y) + \varepsilon_a,$$

where ε_a^α, ε_a^β and ε_a are given real numbers.

Some special examples of uncertainty functions one can also derive from the literature. In [162] strict and weak indiscernibility relations were considered which can define some kind of uncertainty functions. In some cases, it is natural to consider relations defined by so-called ε-indiscernibility [72]. The global uncertainty function for a set of attributes A is usually defined as the intersection i.e. $I_A(x) = \bigcap_{a \in A} I_a(x)$. For some other examples of local and global uncertainty functions see [186].

Different methods of searching for parameters of proper uncertainty functions are discussed for example in [72], [196]. In [196] genetic algorithm was applied for searching for adequate uncertainty functions of the type $I_a^{\varepsilon_a}$, where $\varepsilon_a = f_a(a(x), a(y))$ for every $x, y \in U$. The optimization criterion can be based for example on maximization of the function which combines three quantities. The first quantity can be based on the quality of approximation of classification $\gamma(AS, \{X_1, \ldots, X_r\})$. For specification of the second quantity we first define two relations:

$$R_d = \{(x, y) \in U \times U : d(x) = d(y)\},$$

$$R_{I_A} = \{(x, y) \in U \times U : y \in I_A(x)\}.$$

Using introduced rough inclusions of relations one can measure degree of inclusion of R_d in R_{I_A} in at least three ways:

- $\nu_{SRI}(R_d, R_{I_A})$,
- $\nu_{\pi_1}(R_d, R_{I_A})$,
- $\nu_{\pi_2}(R_d, R_{I_A})$.

The third quantity can be based on the result of the cross-validation test. The idea of cross-validation is depicted on Figure 2.6. For example, in 3–fold cross-validation, the original data table is partitioned into three subtables. From three subtables, a single subtable is retained for testing and the remaining two subtables are used as training data. The cross-validation process is then repeated three times, with each of the three subtables used exactly once as the testing data. The three results from the folds then can be averaged (or otherwise combined) to produce a single estimation.

In the simplest case one can optimize the combination of three quantities as follows:

$$weight_\gamma * \gamma(AS, \{X_1, \ldots, X_r\}) + weight_\nu * \nu_{SRI}(R_d, R_{I_A}) + weight_{test} * test(AS),$$

where
$weight_\gamma, weight_\nu, weight_{test} \geq 0$ and $weight_\gamma + weight_\nu + weight_{test} = 1$.

In the case of $weight_{test} = 0$ first part of the objective function is introduced to prevent shrinking of the positive region of partition. The second part of the objective function is responsible for an increase in the number of connections. We inherit the notion of connections from simple observation, that if $y \in I_A(x)$ then we can say that there is a connection between x and y. We propose to discern two kinds of connections between objects, namely "good" and "bad":

- x has a good connection with y if and only if $(x, y) \in R_{I_A}$ and $(x, y) \in R_d$,
- x has a bad connection with y if and only if $(x, y) \in R_{I_A}$ and $(x, y) \notin R_d$.

We are interested only in connections between objects with the same decision ("good" connections). So, the objective function tries to find out some kind

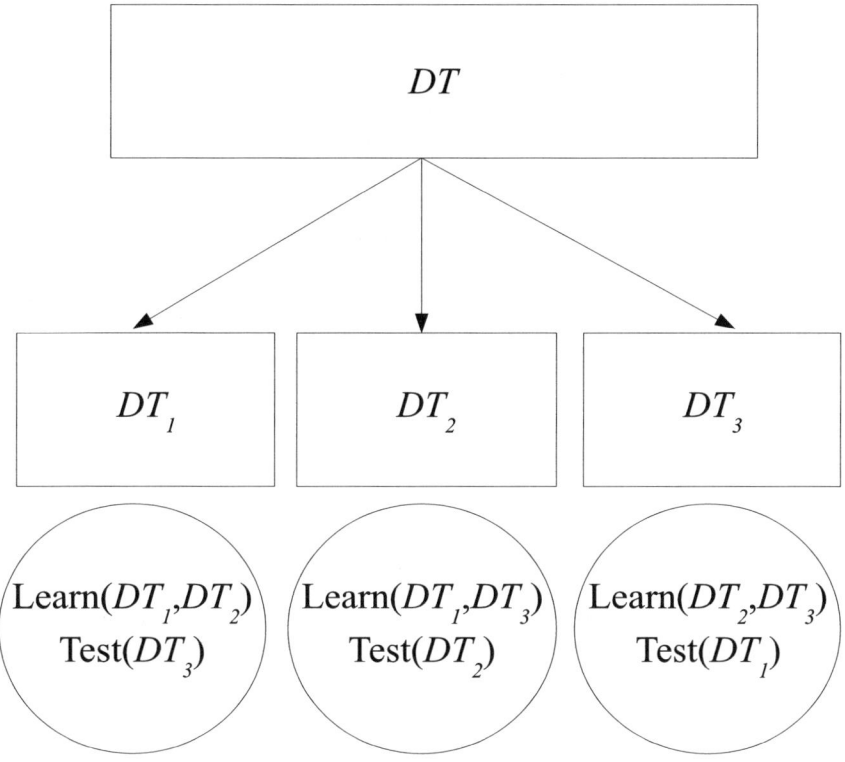

Fig. 2.6. Cross–Validation (CV–3)

of balance between enlarging R_{I_A} and preventing the shrinking of the positive region $POS\left(AS, \{X_1, \ldots, X_r\}\right)$. If we decrease the value of ε_a then for every object $x \in U$ the $I_a^{\varepsilon_a}(x)$ will not change or become larger. So, starting from $\varepsilon_a = 0$ and increasing the value of threshold we can use above property to find all values when $I_a^{\varepsilon_a}(x)$ changes. We can create lists of such thresholds ε_a for each $a \in A$. Next we can check all possible combinations of thresholds to find out the best for our purpose. Of course it can be a long process because, in the worst case, the number of combinations is equal to: $\frac{1}{2} \prod_{a \in A} \left(card\left(V_a\right)^2 - card\left(V_a\right)\right) + 1$. So, it is visible that we need some heuristics to find, maybe not the best of all, but a close to optimal solution in reasonable time. In [196], we use genetic algorithms for this purpose. One application of this method to handwritten numerals recognition is reported in [67].

2.8   Rough Sets in Concept Approximation

The concept approximation problem is the basic problem investigated in machine learning, pattern recognition [47] and data mining [69]. It is necessary to induce approximations of concepts (models of concepts) from available experimental data. The data models developed so far in such areas as statistical learning, machine learning, pattern recognition are not satisfactory for approximation of complex concepts that occur in the perception process. Researchers from different areas have recognized the necessity to work on new methods for concept approximation (see, e.g., [18, 208]). The main reason for this is that these complex concepts are, in a sense, too far from measurements which renders the searching for relevant features in a huge feature space infeasible. There are several research directions aiming at overcoming this difficulty. One of them is based on the interdisciplinary research where the knowledge pertaining to perception in psychology or neuroscience is used to help dealing with complex concepts (see, e.g., [117]). There is a great effort in neuroscience towards understanding the hierarchical structures of neural networks in living organisms [33, 117]. Also mathematicians are recognizing problems of learning as the main problem of the current century [117]. These problems are closely related to complex system modeling as well. In such systems again the problem of concept approximation and its role in reasoning about perceptions is one of the challenges nowadays. One should take into account that modeling complex phenomena entails the use of local models (captured by local agents, if one would like to use the multi-agent terminology [78]) that should be fused afterwards. This process involves negotiations between agents [78] to resolve contradictions and conflicts in local modeling. This kind of modeling is becoming more and more important in dealing with complex real-life phenomena which we are unable to model using traditional analytical approaches. The latter approaches lead to exact models. However, the necessary assumptions used to develop them result in solutions that are too far from reality to be accepted. New methods or even a new science therefore should be developed for such modeling [40].

One of the possible approaches in developing methods for complex concept approximations can be based on the layered learning [200]. Inducing concept approximation should be developed hierarchically starting from concepts that can be directly approximated using sensor measurements toward complex target concepts related to perception. This general idea can be realized using additional domain knowledge represented in natural language. For example, one can use some rules of behavior on the roads, expressed in natural language, to assess from recordings (made, e.g., by camera and other sensors) of actual traffic situations, if a particular situation is safe or not [97]. To deal with such problems one should develop methods for concept approximations together with methods aiming at approximation of reasoning schemes (over such concepts) expressed in natural language. The foundations of such an approach, creating a core of perception logic, are based on rough set theory [106] and its extension rough mereology [104, 122], both invented in Poland, in combination with other soft computing tools, in particular with fuzzy sets.

The outlined problems are some special problems which can be formulated in a more general setting in granular computing.

Information granulation can be viewed as a human way of achieving data compression and it plays a key role in implementing the divide-and-conquer strategy in human problem-solving [226]. Granules are obtained in the process of information granulation.

Granular computing (GC) is based on processing of complex information entities called granules. Generally speaking, granules are collections of entities that are arranged together due to their similarity, functional adjacency or indistinguishability [49, 220, 222, 226].

One of the main branch of GC is Computing with Words and Perceptions (CWP). GC "derives from the fact that it opens the door to computation and reasoning with information which is perception - rather than measurement - based. Perceptions play a key role in human cognition, and underlie the remarkable human capability to perform a wide variety of physical and mental tasks without any measurements and any computations. Everyday examples of such tasks are driving a car in city traffic, playing tennis and summarizing a story" [225, 226].

We consider the optimization tasks in which we are searching for optimal solutions satisfying some constraints. These constraints are often vague, imprecise, and/or specifications of concepts and their dependencies which constitute the constraints, are incomplete. Decision tables [106] are examples of such constraints. Another example of constraints can be found, e.g., [8, 9, 10, 154] where a specification is given by a domain knowledge and data sets. Domain knowledge is represented by an ontology of vague concepts and the dependencies between them. In a more general case, the constraints can be specified in a simplified fragment of a natural language [226].

Granules are constructed using information calculi. Granules are objects constructed in computations aiming at solving the mentioned above optimization tasks. In our approach, we use the general optimization criterion based on the minimal length principle [127, 128]. In searching for (sub-)optimal solutions it is necessary to construct many compound granules using some specific operations such as generalization, specification or fusion. Granules are labeled by parameters. By tuning these parameters we optimize the granules relative to their description size and the quality of data description, i.e., two basic components on which the optimization measures are defined.

From this general description of tasks in GC it follows that together with specification of elementary granules and operation on them it is necessary to define measures of granule quality (e.g., measures of their inclusion, covering or closeness) and tools for measuring the size of granules. Very important are also optimization strategies of already constructed (parameterized) granules.

We consider the above mentioned problems in further chapters of this book.

3 Data Reduction

3.1 Introduction

Nowadays, we deal with large data tables that include up to billions of objects and up to several thousands of attributes. We often face a question whether we can remove some data from a data table preserving its basic properties, that is – whether a table contains some superfluous data. This chapter provides an introduction to rough set based data preprocessing methods, which are concerned with selection of attributes to reduce the dimensionality and improve the data for subsequent data mining analysis.

One of the problems related to practical applications of rough set methods is whether the whole set of attributes is necessary and if not, how to determine the simplified and still sufficient subset of attributes equivalent to the original. The rejected attributes are redundant since their removal cannot worsen the classification. There are usually several such subsets of attributes and those which are minimal with respect to inclusion are called reducts. Finding a minimal reduct (i.e. reduct with a minimal cardinality of attributes among all reducts) is NP-hard [137]. One can also show that the number of reducts of an information system with m attributes may be as large as $\binom{m}{\lfloor m/2 \rfloor}$ (see also Table 3.1).

Table 3.1. Possible Number of Reducts

m	10	12	14	16	18	20	22	24	26	28
$\binom{m}{\lfloor m/2 \rfloor}$	252	924	3432	12870	48620	184756	705432	2704156	10400600	40116600

This means that computation of reducts is a non-trivial task and it can not be solved by a simple increase of computational resources. Significant results in

J. Stepaniuk: Rough - Gran. Comput. in Knowl. Dis. & Data Min., SCI 152, pp. 43–56, 2008.
springerlink.com   © Springer-Verlag Berlin Heidelberg 2008

this area have been achieved in [137]. The problem of finding reducts can be transformed to the problem of finding prime implicants of a monotone Boolean function. An implicant of a Boolean function g is any conjunction of literals (variables or their negations) such that, if the values of these literals are true under an arbitrary valuation val of variables, then the value of the function g under val is also true. A prime implicant is a minimal implicant (with respect to the number of literals). Here we are interested in implicants of monotone Boolean functions i.e. functions constructed without the use of negation. The problem of finding prime implicants is known to be NP-hard, but many heuristics are presented for the computation of one prime implicant.

The general scheme of Boolean reasoning is depicted on Figure 3.1.

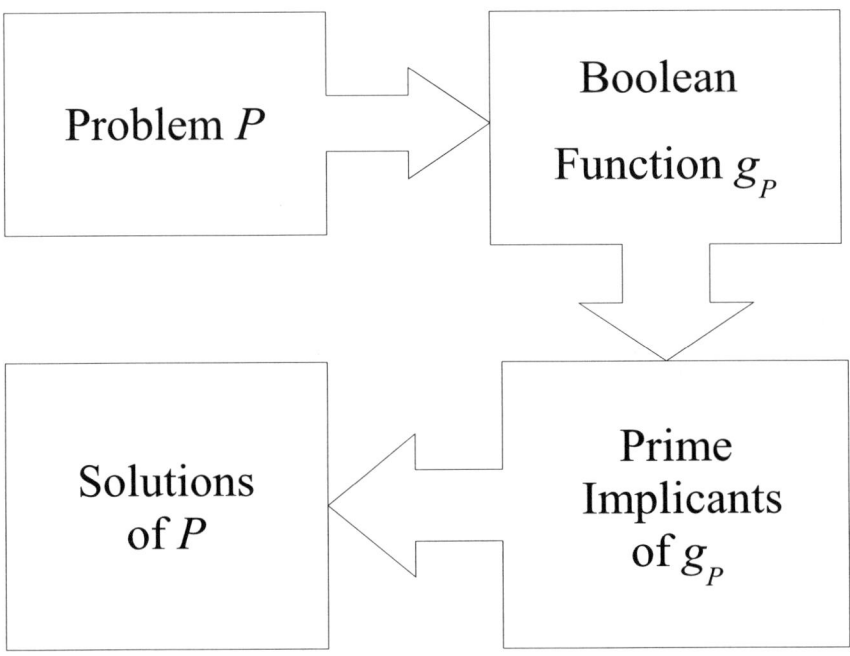

Fig. 3.1. Boolean Reasoning

The general scheme of applying Boolean reasoning to a problem RED of reducts computation can be formulated as follows:

1. Encode the problem RED as a Boolean function g_{RED}.
2. Compute the prime implicants of g_{RED}.
3. Solutions to RED are obtained by interpreting the prime implicants of g_{RED}.

In this chapter we discuss reducts in tolerance rough set model which is a very natural extension of standard rough set model. We consider an approximation spaces AS of the form $AS = (U, I, \nu_{SRI})$ with two conditions for an uncertainty function I :

1. For every $x \in U$, we have $x \in I(x)$ (called reflexivity).
2. For every $x, y \in U$, if $y \in I(x)$, then $x \in I(y)$ (called symmetry).

The chapter is organized as follows. In Section 3.2 the equivalence between reducts and prime implicants of monotonic Boolean functions is investigated. The significance of attributes and the stability of reducts are also considered. In Section 3.3 selection of representative objects is investigated.

3.2   Reducts

In this section we discuss different definitions of a reduct for a single object and for all objects in a given information system [145]. We also propose similar definitions for a decision table. Those definitions have a property that, like in the standard rough set model (see [106, 137]) and in the variable precision rough set model (see [73, 229]) the set of prime implicants of a corresponding discernibility function is equivalent to the set of reducts.

The computation of all types of reducts is based on generalized discernibility matrix. Discernibility matrix was introduced in [137]. We consider dissimilarity instead of discernibility.

Definition 3.1. *Let $IS = (U, A)$ be an information system. By the generalized discernibility matrix we mean the square matrix $(c_{x,y})_{x,y \in U}$, where*

$$c_{x,y} = \{a \in A : y \notin I_a(x)\}.$$

3.2.1   Information Systems and Reducts

Reduct computation can be translated to computing prime implicants of a Boolean function. The type of reduct controls how the Boolean function is constructed.

In the case of reducts for an information system, the minimal sets of attributes that preserve dissimilarity between objects. Thus the full tolerance relation is considered. Therefore resulting reduct is a minimal set of attributes that enables one to introduce the same tolerance relation on the universe as the whole set of attributes does.

In the case of object-related reducts we consider the dissimilarity relation relative to each object. For each object, there are determined the minimal sets of attributes that preserve dissimilarity of that object from all the others. Thus, we construct a Boolean function by restricting the conjunction to only run over the row corresponding to a particular object x of the discernibility matrix (instead of over all rows). Hence, we obtain the discernibility function related to object x. The set of all prime implicants of this function determines the set of reducts of A related to the object x. These reducts reveal the minimum amount of information needed to preserve dissimilarity of x from all other objects.

Let $IS = (U, A)$ be an information system such that the set $A = \{a_1, \ldots, a_m\}$. We assume that $a_1^*, \ldots, a_m^*$ are Boolean variables corresponding to attributes $a_1, \ldots, a_m$, respectively.

In the following definitions we present more formally notions of both types of reducts.

Definition 3.2. *A subset $B \subseteq A$ is called a reduct of A for an object $x \in U$ if and only if*

1. $I_B(x) = I_A(x)$.
2. *For every proper subset $C \subset B$ the first condition is not satisfied.*

Definition 3.3. *A subset $B \subseteq A$ is called a reduct of A if and only if*

1. *For every $x \in U$, we have $I_B(x) = I_A(x)$.*
2. *For every proper subset $C \subset B$ the first condition is not satisfied.*

In the following theorems, we present equivalence between reducts and prime implicants of suitable Boolean functions called discernibility functions.

Theorem 3.4. *For every object $x \in U$ we define the following Boolean function*

$$g_{A,x}(a_1^*, \ldots, a_m^*) = \bigwedge_{y \in U} \bigvee_{a \in c_{x,y}} a^*.$$

The following conditions are equivalent:

1. $\{a_{i_1}, \ldots, a_{i_k}\}$ *is a reduct for the object $x \in U$ in the information system (U, A).*
2. $a_{i_1}^* \wedge \ldots \wedge a_{i_k}^*$ *is a prime implicant of the Boolean function $g_{A,x}$.*

Theorem 3.5. *We define the following Boolean function*

$$g_A(a_1^*, \ldots, a_m^*) = \bigwedge_{x,y \in U} \bigvee_{a \in c_{x,y}} a^*.$$

The following conditions are equivalent:

1. $\{a_{i_1}, \ldots, a_{i_k}\}$ *is a reduct for the information system (U, A).*
2. $a_{i_1}^* \wedge \ldots \wedge a_{i_k}^*$ *is a prime implicant of the Boolean function g_A.*

Example 3.6. The flags data table was adopted from the book [131]. The data set is a table listing various features of the flags of different states of the USA, along with the information whether or not the state was a union (U) or confederate (C) state during the civil war. For simplicity of presentation we only consider part of this data set, namely $DT = (U, A \cup \{d\})$, where $U = \{x_1, \ldots x_9\}$ and $A = \{a_1, a_2, a_3, a_4\}$ (see Table 3.2).

For presentation of reducts in the information system (U, A), we assume that the last column ("Type") is not given.

Table 3.2. Flags Data Table

		a_1	a_2	a_3	a_4	d
Flag		Stars	Hues	Number	Word	Type
Alabama	x_1	0	2	0	0	C
Virginia	x_2	0	5	0	4	C
Tennesee	x_3	3	3	0	0	C
Texas	x_4	1	3	0	0	C
Louisiana	x_5	0	4	0	4	C
Illinois	x_6	0	6	2	6	U
Iowa	x_7	0	5	0	10	U
Ohio	x_8	17	3	0	0	U
New Jersey	x_9	0	5	1	3	U

Table 3.3. Discernibility Matrix for Information System

	x_1	x_2	x_3	x_4	x_5	x_6	x_7	x_8	x_9
x_1	—	a_2	—	—	—	a_2, a_3	a_2, a_4	a_1	a_2
x_2	a_2	—	—	—	—	a_3	—	a_1	—
x_3	—	—	—	—	—	a_2, a_3	a_4	a_1	—
x_4	—	—	—	—	—	a_2, a_3	a_4	a_1	—
x_5	—	—	—	—	—	a_3	—	a_1	—
x_6	a_2, a_3	a_3	a_2, a_3	a_2, a_3	a_3	—	a_3	a_1, a_2, a_3	—
x_7	a_2, a_4	—	a_4	a_4	—	a_3	—	a_1, a_4	—
x_8	a_1	a_1	a_1	a_1	a_1	a_1, a_2, a_3	a_1, a_4	—	a_1
x_9	a_2	—	—	—	—	—	—	a_1	—

We consider for all attributes $a \in A$, the following uncertainty function:

$$y \in I_a^{\varepsilon_a}(x) \quad \text{if and only if } diff_{f_a}(a(x), a(y)) \leq \varepsilon_a \quad \text{i.e.} \quad \frac{|a(x) - a(y)|}{\max_a - \min_a} \leq \varepsilon_a.$$

We choose the following thresholds:

$\varepsilon_{a_1} = \frac{5}{17}$, $\varepsilon_{a_2} = 0.5$, $\varepsilon_{a_3} = 0.5$ and $\varepsilon_{a_4} = 0.7$.

We consider an approximation space $AS_A = (U, I_A, \nu_{SRI})$, where the global uncertainty function $I_A : U \to P(U)$ is defined by

$$I_A(x) = \bigcap_{a \in A} I_a^{\varepsilon_a}(x).$$

We construct the discernibility matrix (see Table 3.3). For example for the object x_3 we obtain the following discernibility function

$$g_{A,x}(a_1^*, a_2^*, a_3^*, a_4^*) = (a_2^* \vee a_3^*) \wedge a_4^* \wedge a_1^*.$$

Object related reducts are presented in Table 3.4.

For the information system we can not reduce the set of attributes, namely there is only one reduct equal to $\{a_1, a_2, a_3, a_4\}$.

Table 3.4. Object Related Reducts in Information System

	Object related reducts
x_1	$\{a_1, a_2\}$
x_2	$\{a_1, a_2, a_3\}$
x_3	$\{a_1, a_2, a_4\}, \{a_1, a_3, a_4\}$
x_4	$\{a_1, a_2, a_4\}, \{a_1, a_3, a_4\}$
x_5	$\{a_1, a_3\}$
x_6	$\{a_3\}$
x_7	$\{a_3, a_4\}$
x_8	$\{a_1\}$
x_9	$\{a_1, a_2\}$

3.2.2   Decision Tables and Reducts

We present methods of attributes set reduction in a decision table. General scheme of this approach is depicted on Figure 3.2.

If we consider a decision table instead of an information system, this translates to a modification of the discernibility function constructed for the information system. Since we do not need to preserve dissimilarity between objects with the same decision, we can delete those expressions from the discernibility function that preserve dissimilarity between objects within the same decision class. Thus, a resulting reduct is a minimal set of attributes that enables to make the same decisions as the whole set of attributes.

In the case of computing reducts in a decision table, decision rules can be also computed at the same time for efficiency reasons. Let us note that object-related reducts will typically produce shorter rules than reducts calculated for decision table, where length of rules is measured in the number of selectors used in an induced rule.

Let $DT = (U, A \cup \{d\})$ be a decision table. In the following definitions we present more formally all types of reducts.

Definition 3.7. *A subset $B \subseteq A$ is called a relative reduct of A for an object $x \in U$ if and only if*

1. $\{y \in I_B(x) : d(y) \neq d(x)\} = \{y \in I_A(x) : d(y) \neq d(x)\}$.
2. For every proper subset $C \subset B$ the first condition is not satisfied.

Illustrative example is depicted on Figure 3.3.

Definition 3.8. *A subset $B \subseteq A$ is called a relative reduct of A if and only if*

1. $POS(AS_B, \{d\}) = POS(AS_A, \{d\})$.
2. For every proper subset $C \subset B$ the first condition is not satisfied.

In the following theorems we demonstrate an equivalence between relative reducts and prime implicants of suitable Boolean functions.

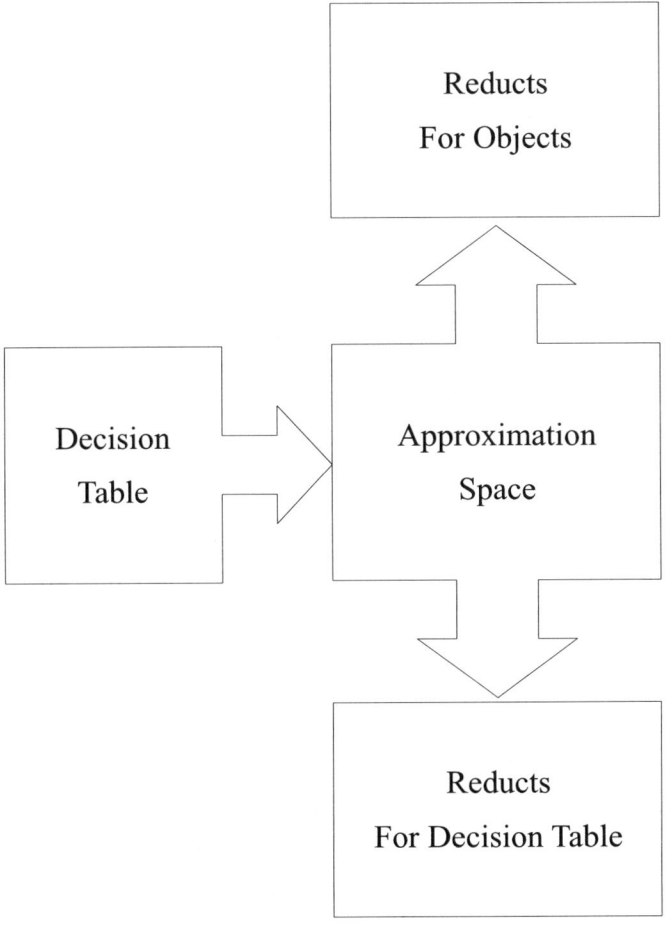

Fig. 3.2. Two Kinds of Reducts in Decision Table

Theorem 3.9. *For every object $x \in U$ and the Boolean function defined by*

$$g_{A \cup \{d\}, x}(a_1^*, \ldots, a_m^*) = \bigwedge_{y \in U, d(y) \neq d(x)} \bigvee_{a \in c_{x,y}} a^*$$

the following conditions are equivalent:

1. $\{a_{i_1}, \ldots, a_{i_k}\}$ *is a relative reduct for the object $x \in U$ in the decision table DT.*
2. $a_{i_1}^* \wedge \ldots \wedge a_{i_k}^*$ *is a prime implicant of the Boolean function $g_{A \cup \{d\}, x}$.*

Theorem 3.10. *For the Boolean function defined by*

$$g_{A \cup \{d\}}(a_1^*, \ldots, a_m^*) = \bigwedge_{x, y \in U, d(y) \neq d(x)} \bigvee_{a \in c_{x,y}} a^*$$

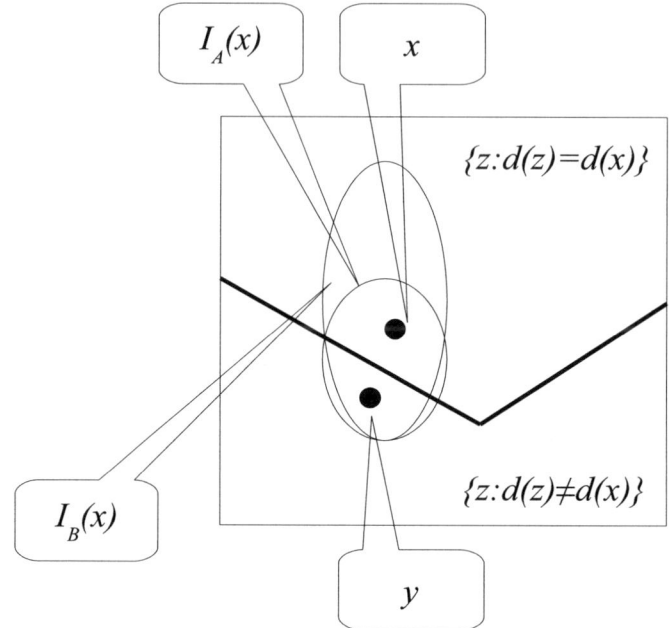

Fig. 3.3. Idea of Object Related Reducts

the following conditions are equivalent:

1. $\{a_{i_1}, \ldots, a_{i_k}\}$ *is a relative reduct of* A.
2. $a_{i_1}^* \wedge \ldots \wedge a_{i_k}^*$ *is a prime implicant of the Boolean function* $g_{A \cup \{d\}}$.

Example 3.11. We consider Table 3.2 with decision attribute "Type". We obtain the relative (with respect to decision) discernibility matrix described in Table 3.5. For example for object x_7 the discernibility function

$$g_{A \cup \{d\}, x_7} (a_1^*, a_2^*, a_3^*, a_4^*) = (a_2^* \vee a_4^*) \wedge a_4^* \wedge a_4^* \equiv a_4^*.$$

In Table 3.6 we present the set of all relative reducts related to particular objects.

Now we discuss a heuristic [196] which can be applied to the computation of relative reducts without explicit use of discernibility function. We can also use the presented method to simplification of discernibility function.

To find one relative reduct, we build a discernibility matrix. Next, we make reduction of superfluous entries in this matrix. We set an entry to be empty if it is a superset of another non-empty entry. At the end of this process we obtain the set $COMP$ of so called components. From the set of components the described type of reduct can be generated by applying Boolean reasoning. We present heuristics for computing one reduct of the considered type with the possibly minimal number of attributes. These heuristics can produce sets which

Table 3.5. Relative Discernibility Matrix

	x_1	x_2	x_3	x_4	x_5	x_6	x_7	x_8	x_9
x_1	$-$	$-$	$-$	$-$	$-$	a_2, a_3	a_2, a_4	a_1	a_2
x_2	$-$	$-$	$-$	$-$	$-$	a_3	$-$	a_1	$-$
x_3	$-$	$-$	$-$	$-$	$-$	a_2, a_3	a_4	a_1	$-$
x_4	$-$	$-$	$-$	$-$	$-$	a_2, a_3	a_4	a_1	$-$
x_5	$-$	$-$	$-$	$-$	$-$	a_3	$-$	a_1	$-$
x_6	a_2, a_3	a_3	a_2, a_3	a_2, a_3	a_3	$-$	$-$	$-$	$-$
x_7	a_2, a_4	$-$	a_4	a_4	$-$	$-$	$-$	$-$	$-$
x_8	a_1	a_1	a_1	a_1	a_1	$-$	$-$	$-$	$-$
x_9	a_2	$-$	$-$	$-$	$-$	$-$	$-$	$-$	$-$

Table 3.6. Relative Object Related Reducts

	Object related reducts
x_1	$\{a_1, a_2\}$
x_2	$\{a_1, a_3\}$
x_3	$\{a_1, a_2, a_4\}, \{a_1, a_3, a_4\}$
x_4	$\{a_1, a_2, a_4\}, \{a_1, a_3, a_4\}$
x_5	$\{a_1, a_3\}$
x_6	$\{a_3\}$
x_7	$\{a_4\}$
x_8	$\{a_1\}$
x_9	$\{a_2\}$

are supersets of considered reducts but they are much more efficient than the generic procedure.

First, we introduce a notion of a minimal distinction. By a minimal distinction (md, in short) we understand a minimal set of attributes sufficient to discern between two objects. Let us note that the minimal component com consists of minimal distinctions and $card(com)$ is equal or greater than $card(md)$. We say, that md is indispensable if there is a component made of only one md. We include all attributes from the indispensable md to R. Then from $COMP$ we eliminate all these components which have at least one md equal to md in R. It is important that the process of selecting attributes to R will be finished when the set $COMP$ is empty. We calculate for any md from $COMP$:

$$c\,(md) = w_1 * c_1\,(md) + w_2 * c_2\,(md), \text{ where}$$
$$c_1\,(md) = \left(\frac{card(md \cap R)}{card(md)} \right)^p \text{ and}$$
$$c_2\,(md) = \left(\frac{card(\{com \in COMP : \exists_{md' \subset com} md' \subset (R \cup md)\})}{card(COMP)} \right)^q.$$

For example, we can assume $p = q = 1$.

The first function is a "measure of extending" of R. Since we want to minimize cardinality of R, we are interested in finding md with the largest intersection with actual R. In this way, we always add to R an almost minimal number of new attributes. The second measure is used to examine profit by adding attributes from md to R. We want to include in R the most frequent md in $COMP$ and minimize $COMP$ as much as possible. When $c_2(md) = 1$, then after "adding this md" to R we will obtain a pseudo-reduct i.e. a superset of a reduct.

3.2.3   Significance of Attributes and Stability of Reducts

A problem of relevant attribute selection is one of the importance and have been studied in machine learning and knowledge discovery [69]. There are also several attempts to this problem based on rough sets.

One of the first ideas in rough set based attribute selection [106] was to consider as relevant those attributes which are in the *core* of an information system, i.e. attributes that belong to the intersection of all reducts of that system.

It is also possible to consider as relevant attributes those from some approximate reducts of sufficiently high quality. As it follows from the considerations concerning reduction of attributes, they can not be equally important and some of them can be eliminated from a data table without loose information contained in the table. The idea of attribute reduction can be generalized by introduction of the concept of *attribute significance*. This concept enables evaluation of attributes not only over a two-valued scale, *dispensable – indispensable*, but by associating with an attribute a real number from the $[0, 1]$ closed interval. This number expresses the relevance of the attribute in the information table.

Significance of an attribute $a \in B \subseteq A$ in a decision table $DT = (U, A \cup \{d\})$ can be evaluated by measuring the effect of removing of an attribute $a \in B$ from the attribute set B on the positive region defined by the table DT. As shown previously, the number $\gamma(AS_B, \{d\})$ expresses the degree of dependency between attributes B and d. We can ask how the coefficient $\gamma(AS_B, \{d\})$ changes when an attribute a is removed, i.e., what is the difference between $\gamma(AS_B, \{d\})$ and $\gamma(AS_{B-\{a\}}, \{d\})$. We can normalize the difference and define the significance of an attribute a as

$$\sigma_{(AS_B, \{d\})}(a) = \frac{\gamma(AS_B, \{d\}) - \gamma(AS_{B-\{a\}}, \{d\})}{\gamma(AS_B, \{d\})}.$$

Thus the coefficient $\sigma(a)$ can be understood as the error of classification which occurs when attribute a is dropped.

Example 3.12. We consider Table 3.2 with decision attribute "Type". Let $AS_A = (U, I_A, \nu_{SRI})$ be an approximation space defined in Example 3.6. The uncertainty function I_A is presented in Table 3.7.

We obtain

$$LOW(AS_A, \|d = C\|_{DT}) = \{x_1\},$$

$$LOW(AS_A, \|d = U\|_{DT}) = \{x_6, x_8\},$$

Table 3.7. Uncertainty Function

	$I_A(\bullet)$
x_1	$\{x_1, x_3, x_4, x_5\}$
x_2	$\{x_2, x_3, x_4, x_5, x_7, x_9\}$
x_3	$\{x_1, x_2, x_3, x_4, x_5, x_9\}$
x_4	$\{x_1, x_2, x_3, x_4, x_5, x_9\}$
x_5	$\{x_1, x_2, x_3, x_4, x_5, x_7, x_9\}$
x_6	$\{x_6, x_9\}$
x_7	$\{x_2, x_5, x_7, x_9\}$
x_8	$\{x_8\}$
x_9	$\{x_2, x_3, x_4, x_5, x_6, x_7, x_9\}$

hence $\gamma(AS_A, \{d\}) = \frac{3}{9}$. For every attribute $a \in A$ we obtain:

$$\gamma(AS_{A-\{a_1\}}, \{d\}) = \frac{1}{9} \quad \text{and} \quad \sigma_{(AS_A, \{d\})}(a_1) = \frac{2}{3},$$

$$\gamma(AS_{A-\{a_2\}}, \{d\}) = \frac{2}{9} \quad \text{and} \quad \sigma_{(AS_A, \{d\})}(a_2) = \frac{1}{3},$$

$$\gamma(AS_{A-\{a_3\}}, \{d\}) = \frac{2}{9} \quad \text{and} \quad \sigma_{(AS_A, \{d\})}(a_3) = \frac{1}{3},$$

$$\gamma(AS_{A-\{a_4\}}, \{d\}) = \frac{3}{9} \quad \text{and} \quad \sigma_{(AS_A, \{d\})}(a_4) = 0.$$

Another approach to the problem of relevant attributes selection is related to dynamic reducts (see e.g. [5]) i.e. condition attribute sets appearing "sufficiently often" as reducts of samples of the original decision table. The attributes belonging to the "majority" of dynamic reducts are defined as relevant. The value thresholds for "sufficiently often" and "majority" need to be tuned for the given data. Several of the reported experiments show that the set of decision rules based on such attributes is much smaller than the set of all decision rules and the quality of classification of new objects is increasing or at least not significantly decreasing if only rules constructed over such relevant attributes are considered.

We recall the notion of a stability coefficient of a reduct [5]. Let $DT = (U, A \cup \{d\})$ be a decision table. One can say that $DT' = (U', A \cup \{d\})$ is a subtable of DT if and only if $U' \subset U$. Let $B \subseteq A$ be a relative reduct for $(AS_A, \{d\})$. Let F denote a set of subtables of $DT = (U, A \cup \{d\})$. The number

$$\frac{card\left(\{DT' \in F : B \text{ is a reduct in } DT'\}\right)}{card\left(F\right)}$$

is called the stability coefficient of the reduct B.

3.3 Representatives

In some sense dual to the problem of attribute set reduction is the problem of object number reduction (selection). In the standard rough set approach it seems that the first idea is to take one element from every equivalence class defined by a set of attributes. When we consider overlapping classes the above idea should be modified. In this section we discuss the problem of proper representative object selection from data tables. We discuss equivalence of the problem of object number reduction to the problem of prime implicants computation for a suitable Boolean function.

The general problem can be described as follows:

Given a set of objects U, the reduction process of U consists in finding a new set $U' \subset U$. The objects which belong to the set U' are chosen for example by using an evaluation criterion. The main advantage of the evaluation criterion approach is that a simple evaluation criterion can be defined which ensures a high level of efficiency. On the other hand, the definition of the evaluation criterion is a difficult problem, because in the new data set some objects are dropped and only a good evaluation criterion preserves the effectiveness of the knowledge acquired during the subsequent learning process.

There are many methods of adequate representative selection (see for example [25, 41, 88, 119, 120, 189, 196]).

In the standard rough set model, representatives can be computed from every indiscernibility class. In this section we discuss representative selection based on generalized approximation spaces and Boolean reasoning. This approach was suggested in [136].

3.3.1 Representatives in Information Systems

We assume that $AS = (U, I_A, \nu_{SRI})$ is an approximation space, where $U = \{x_1, \ldots, x_n\}$ is a set of objects and let $x_1^*, \ldots, x_n^*$ be Boolean variables corresponding to objects $x_1, \ldots, x_n$, respectively.

Definition 3.13. *Let (U, A) be an information system. A subset $U' \subseteq U$ is a minimal set of representatives if and only if the following two conditions are satisfied:*

1. *For every $x \in U$ there is $y \in U'$ such that $x \in I_A(y)$.*
2. *For every proper subset $U" \subset U'$ the first condition is not satisfied.*

In the next theorem we obtain a characterization of minimal sets of representatives in information systems.

Theorem 3.14. *We define the Boolean function*

$$g_{(U,A)}(x_1^*, \ldots, x_n^*) = \bigwedge_{x_i \in U} \bigvee_{x_j \in I_A(x_i)} x_j^*.$$

The following conditions are equivalent:

1. $\{x_{i_1}, \ldots, x_{i_k}\}$ *is a minimal set of representatives.*
2. $x_{i_1}^* \wedge \ldots \wedge x_{i_k}^*$ *is a prime implicant of the Boolean function* $g_{(U,A)}$.

Example 3.15. We consider Table 3.2 without decision attribute "Type". From the uncertainty function I_A presented in Table 3.7, we obtain the Boolean function

$$g_{(U,A)}\left(x_1^*, \ldots, x_9^*\right) = \left(x_1^* \vee x_3^* \vee x_4^* \vee x_5^*\right) \wedge \left(x_2^* \vee x_3^* \vee x_4^* \vee x_5^* \vee x_7^* \vee x_9^*\right) \wedge \ldots \wedge \left(x_2^* \vee x_3^* \vee x_4^* \vee x_5^* \vee x_6^* \vee x_7^* \vee x_9^*\right).$$

Computing prime implicants of $g_{(U,A)}$, we obtain the following sets of representatives:

$\{x_1, x_2, x_6, x_8\}$, $\{x_1, x_6, x_7, x_8\}$, $\{x_2, x_3, x_6, x_8\}$, $\{x_3, x_6, x_7, x_8\}$, $\{x_2, x_4, x_6, x_8\}$, $\{x_4, x_6, x_7, x_8\}$, $\{x_5, x_6, x_8\}$, $\{x_1, x_8, x_9\}$, $\{x_3, x_8, x_9\}$, $\{x_4, x_8, x_9\}$ and $\{x_5, x_8, x_9\}$.

3.3.2   Representatives in Decision Tables

In decision tables, we also consider the decision in computation of minimal sets of representatives.

Definition 3.16. *Let* $(U, A \cup \{d\})$ *be a decision table. A subset* $U' \subseteq U$ *is a relative minimal set of representatives if and only if the following two conditions are satisfied:*

1. *For every* $x \in U$ *there is* $y \in U'$ *such that* $x \in I_A(y)$ *and* $d(x) = d(y)$.
2. *For every proper subset* $U'' \subset U'$ *the first condition is not satisfied.*

Illustrative example is depicted in Figure 3.4. We can formulate a similar theorem for computation of representatives in a decision table as with computation of relative reducts.

Theorem 3.17. *Let* $ST(x_i) = \{x_j \in U : x_j \in I_A(x_i), d(x_i) = d(x_j)\}$.
We define the Boolean function

$$g_{(U,A\cup\{d\})}\left(x_1^*, \ldots, x_n^*\right) = \bigwedge_{x_i \in U} \bigvee_{x_j \in ST(x_i)} x_j^*.$$

The following conditions are equivalent:

1. $\{x_{i_1}, \ldots, x_{i_k}\}$ *is a relative minimal set of representatives.*
2. $x_{i_1}^* \wedge \ldots \wedge x_{i_k}^*$ *is a prime implicant of* $g_{(U,A\cup\{d\})}$.

Below we sketch an algorithm for computation of one set of representatives with minimal or near minimal number of elements.

The main difference between finding out one set of representatives and one relative reduct is in the way in which we calculate and interpret components.

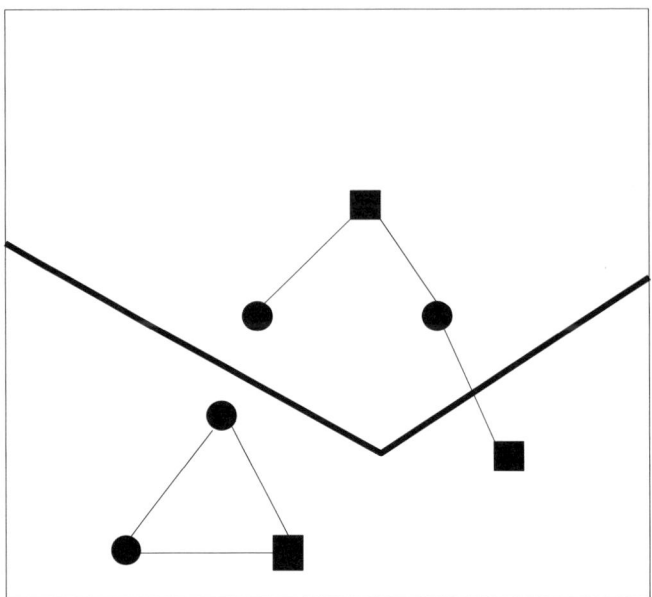

Fig. 3.4. Representatives (Circles Represent Objects and Squares Representatives)

In case of the relative set of representatives we do not build the discernibility matrix, but we replace it by a similar table containing for any object x_i all objects similar to x_i and with the same decision:

$$ST\left(x_i\right) = \left\{x_j \in U : x_j \in I_A\left(x_i\right), d\left(x_i\right) = d\left(x_j\right)\right\}.$$

After reduction, we obtain components as essential entries in ST. For $COMP$ we can apply the algorithm used to compute a reduct assuming $card(md) = 1$. We add to the constructed relative absorbent set any object which is the most frequent in $COMP$ and then eliminate from $COMP$ all components having this object. This process terminates when $COMP$ is empty. For more details see [196].

Part II

Classification and Clustering

4 Selected Classification Methods

Any classification algorithm should consists of some classifiers together with a method of conflicts resolving between the classifiers when new objects are classified. In this chapter we discuss two classes of classification algorithms. Algorithms from the first class are using sets of decision rules as classifiers together with some methods of conflict resolving. The rules are generated from decision tables with tolerance relations using Boolean reasoning approach. They create decision classes descriptions. However, to predict (or classify) new object to a proper decision class it is necessary to fix some methods for conflict resolving between rules recognizing the object and voting for different decisions. We also discuss how such decision rules can be generated using Boolean reasoning. Algorithms of the second kind are based on the nearest neighbor method ($k - NN$) (see, top ten data mining algorithms [217]). We show how this method can be combined with some rough set method for relevant attribute selection.

The organization of this chapter is as follows. In Section 4.1 the concept of granularity in rules is discussed. In Section 4.2 decision rules in standard and tolerance rough set model are discussed. In Section 4.3 we give some overview of quantitative measures for decision rule ranking. The received ranking can be used for rule filtration or for conflict between decision rules resolving when new objects are classified. In Section 4.4 we discuss a hybrid method received by combining the $k - NN$ method with some methods for relevant attribute selection.

4.1 Information Granulation and Rules

We present some definitions concerning decision rules also called classification rules. A rule

If λ then ξ

is composed of a condition λ and a decision part ξ. The decision part usually describes the predicted class. The condition part λ is a conjunction of selectors, each selector being a condition involving a single attribute. For nominal attributes, this condition is a simple equality test, for example $Sex = f$. For

J. Stepaniuk: Rough - Gran. Comput. in Knowl. Dis. & Data Min., SCI 152, pp. 59–66, 2008.
springerlink.com

numerical attributes, the condition is typically inclusion in an interval, for example $7 \leq Age < 13$. Decision rules for nominal attributes are represented as statements in the following form:

If $a_1 = v_1$ **and** ... **and** $a_k = v_k$ **then** $d = i$.

To emphasize the use of information granulation in rule-based computing [21], let us consider a rule of the form

If λ **then** ξ

where λ and ξ are represented as numerical intervals in the space of real valued attributes. A low level of granularity of the condition λ associated with a high level of granularity of the conclusion ξ describes a rule of high relevance: it applies to a wide range of objects (as its condition λ is not very detailed) while offering a very specific conclusion ξ. On the other hand, if we encounter a rule containing a very detailed condition λ with quite limited applicability while the conclusion ξ is quite general, we may view rule's relevance to be quite limited. In general, increasing granularity of the condition λ and decreasing granularity of conclusion ξ decrease the quality of the rule.

4.2   Decision Rules in Rough Set Models

In this section we show how to use Boolean reasoning approach for decision rule generation from data tables extended by tolerance relations defined on the attribute value vectors. The general scheme is depicted on Figure 4.1.

It is important to note that the aim of Boolean reasoning in considered problems is to preserve some properties described by discernibility relations. In the classical rough set approach the discernibility relation is equal to the complement of the indiscernibility relation. However, in more general cases, (e.g. related to tolerance relation) one can define the discernibility relation in a quite different way (e.g. one can define the discernibility not by $I(x) \cap I(y) = \emptyset$, not by $y \notin I(x)$). We discuss how the Boolean reasoning method works for the classical case of discernibility relation $\{(x, y) \in U \times U : y \notin I(x)\}$. However, one can easily modify it for other cases of discernibility relations. The decision rule generation process for, so called minimal rules, can be reduced to the generation of relative object related reducts. These reducts can be computed as prime implicants of an appropriate Boolean function. Let us mention that the presented methods can be extended for computing of approximate decision rules e.g. association rules [2, 95].

Decision rules are generated from reducts. So in order to compute decision rules, reducts have to be computed first. One can use different kinds of reducts (object-related or for data table) and different methods of reducts computation. For example the reducts one can compute by exhaustive calculation. This method finds all reducts by computing prime implicants of a Boolean function.

For example one can use object-related reducts. One can conceptually overlay each reduct over the decision table it was computed from, and read off the

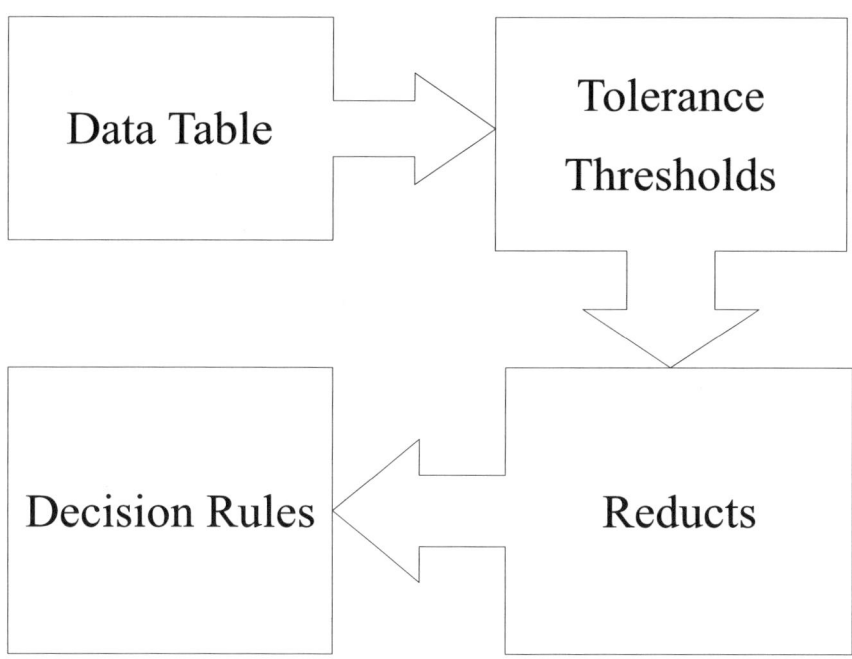

Fig. 4.1. Decision Rules in Tolerance Rough Set Model

attribute values. We give two examples of selectors with respect to two different types of uncertainty functions:

If an uncertainty function I_a for an attribute $a \in A$ is defined by
$y \in I_a(x)$ if and only if $a(x) = a(y)$
(as in the standard rough set model) then for the attribute a from some reduct related to an object $x \in U$ a constructed selector is of the form $a = a(x)$.

If an uncertainty function $I_a^{\varepsilon_a}$ for a numeric attribute $a \in A$ is defined by
$y \in I_a^{\varepsilon_a}(x)$ if and only if $\delta_a(a(x), a(y)) \leq \varepsilon_a$,
where δ_a is a distance function and $\varepsilon_a \in [0, 1]$ is a real number then

for the attribute a from some relative reduct related to an object $x \in U$ a constructed selector should exploit the form of a distance function. For example, for the distance function $diff_a$ a constructed selector is of the form

$$a \in [a(x) - \varepsilon_a * (max_a - min_a), a(x) + \varepsilon_a * (max_a - min_a)].$$

Example 4.1. Let us consider data from Table 3.2. Using relative object related reducts (see Table 3.6) we obtain rules presented in Table 4.1.

In [132] and [133] tolerance rough set model to approximate rule calculation is discussed. A genetic algorithm is used for fuzzification of obtained decision rules. Presented experimental results show that the proposed method allows getting a smaller set of decision rules with usually better classification abilities.

Table 4.1. Rules Based on Relative Object Related Reducts

Objects	Rules
x_1	if $a_1 \in [0,5]$ and $a_2 \in [2,4]$ then $d = C$
x_2	if $a_1 \in [0,5]$ and $a_3 \in [0,1]$ then $d = C$
x_3	if $a_1 \in [0,8]$ and $a_2 \in [2,5]$ and $a_4 \in [0,7]$ then $d = C$
x_3	if $a_1 \in [0,8]$ and $a_3 \in [0,1]$ and $a_4 \in [0,7]$ then $d = C$
x_4	if $a_1 \in [0,6]$ and $a_2 \in [2,5]$ and $a_4 \in [0,7]$ then $d = C$
x_4	if $a_1 \in [0,6]$ and $a_3 \in [0,1]$ and $a_4 \in [0,7]$ then $d = C$
x_5	if $a_1 \in [0,5]$ and $a_3 \in [0,1]$ then $d = C$
x_6	if $a_3 \in [1,2]$ then $d = U$
x_7	if $a_4 \in [3,10]$ then $d = U$
x_8	if $a_1 \in [12,17]$ then $d = U$
x_9	if $a_2 \in [3,6]$ then $d = U$

4.3   Evaluation of Decision Rules

Decision rules induced from a data table can be evaluated along at least two dimensions: performance (prediction) and explanatory features (description). By performance is meant assessment of how well the set of rules does in classifying new objects, according to some specified performance criterion. By explanatory features is meant how interpretable the rules are, so that one might gain some insight into how the classification or decision making process is carried out. How these two evaluation dimensions are to weighted is a matter of the intended role of the generated rules. If the set of rules is to operate in a fully automated environment, then performance may be the main feature of interest. Conversely, if the set of rules induction is part of a knowledge discovery process, then the interpretability of the rules will be more important.

Different methods for classification vary in how much they facilitate the knowledge discovery aspect, depending on the type of classifiers they produce. A point that is often held forth in favor of methods that produce rule sets is that the models are directly readable and interpretable.

For example, classification one can perform by the following algorithm. Presented with a given object (information vector) to classify, the algorithm scans through the rule set and determines if each rule fires (is applicable). Rules whose antecedent are not in direct conflict with the contents of the information vector fire. If no rules fire, the most frequent decision in the decision table is taken. This means that the dominating decision in the data set is suggested. If more than one rule fires, these may indicate more than one possible decision class. An election process among the firing rules is then performed in order to resolve conflicts and rank the decisions. First, one can accumulate the votes for each possible decision by each firing rule. Second, the accumulated number of votes for each possible decision defines a certainty coefficient for each decision class. The decision class with the largest certainty coefficient is selected. Ties are resolved by the majority voter algorithm.

Table 4.2. Contingency Table Representing the Quantitative Information about the Rule

	ξ	$\neg\xi$	
λ	$card\left(\|\lambda \wedge \xi\|_{DT}\right)$	$card\left(\|\lambda \wedge \neg\xi\|_{DT}\right)$	$card\left(\|\lambda\|_{DT}\right)$
$\neg\lambda$	$card\left(\|\neg\lambda \wedge \xi\|_{DT}\right)$	$card\left(\|\neg\lambda \wedge \neg\xi\|_{DT}\right)$	$card\left(\|\neg\lambda\|_{DT}\right)$
	$card\left(\|\xi\|_{DT}\right)$	$card\left(\|\neg\xi\|_{DT}\right)$	$card\left(U\right)$

Let $DT = (U, A \cup \{d\})$ be a given data table. We use the set-theoretical interpretation of rules. It relates a rule to data sets from which the rule is discovered.

Using the cardinalities of sets, we obtain the 2×2 contingency table representing the quantitative information about the rule **if** λ **then** ξ (see Table 4.2).

Using the elements of the contingency table, we may define the support of a decision rule *Rule* of the form **if** λ **then** ξ by

$$support_{DT}\left(Rule\right) = card\left(\|\lambda\|_{DT} \cap \|\xi\|_{DT}\right)$$

and its accuracy by

$$accuracy_{DT}\left(Rule\right) = \frac{card\left(\|\lambda\|_{DT} \cap \|\xi\|_{DT}\right)}{card\left(\|\lambda\|_{DT}\right)}.$$

This quantity shows the degree to which λ implies ξ. It may be viewed as the conditional probability of a randomly selected object satisfying ξ given that the element satisfies λ. In set-theoretic terms, it is the degree to which $\|\lambda\|_{DT}$ is included in $\|\xi\|_{DT}$ and is equal to $\nu_{SRI}\left(\|\lambda\|_{DT}, \|\xi\|_{DT}\right)$. Different names were given to this measure, including, the confidence (for mining association rules [2]) and the absolute support [219].

The coverage of *Rule* is defined by

$$coverage_{DT}\left(Rule\right) = \frac{card\left(\|\lambda\|_{DT} \cap \|\xi\|_{DT}\right)}{card\left(\|\xi\|_{DT}\right)}.$$

In set-theoretic term, it is the degree to which $\|\xi\|_{DT}$ is included in $\|\lambda\|_{DT}$ and is equal to $\nu_{SRI}\left(\|\xi\|_{DT}, \|\lambda\|_{DT}\right)$.

In Figure 4.2 are depicted four decision rules types for ξ defined by condition $d = 1$. Each dashed set represents $\|\lambda\|_{DT}$.

The rule quality for the rules in a rule set *Rule_Set* is determined by a quality function:

$$q : Rule\_Set \rightarrow [0, 1].$$

Michalski [87] suggests that high accuracy and coverage are requirements of decision rules. q_M is a weighted sum of the measures of the accuracy and the coverage properties:

$$q_M(Rule) = w * accuracy_{DT}(Rule) + (1 - w) * coverage_{DT}(Rule).$$

In the Torgo quality function q_T [203] the accuracy value is judged to be the more important than coverage. The weight is made dependent on accuracy:

$$q_T(Rule) \text{ is } q_M(Rule) \text{ with } w = \frac{1}{2} + \frac{1}{4}accuracy_{DT}(Rule).$$

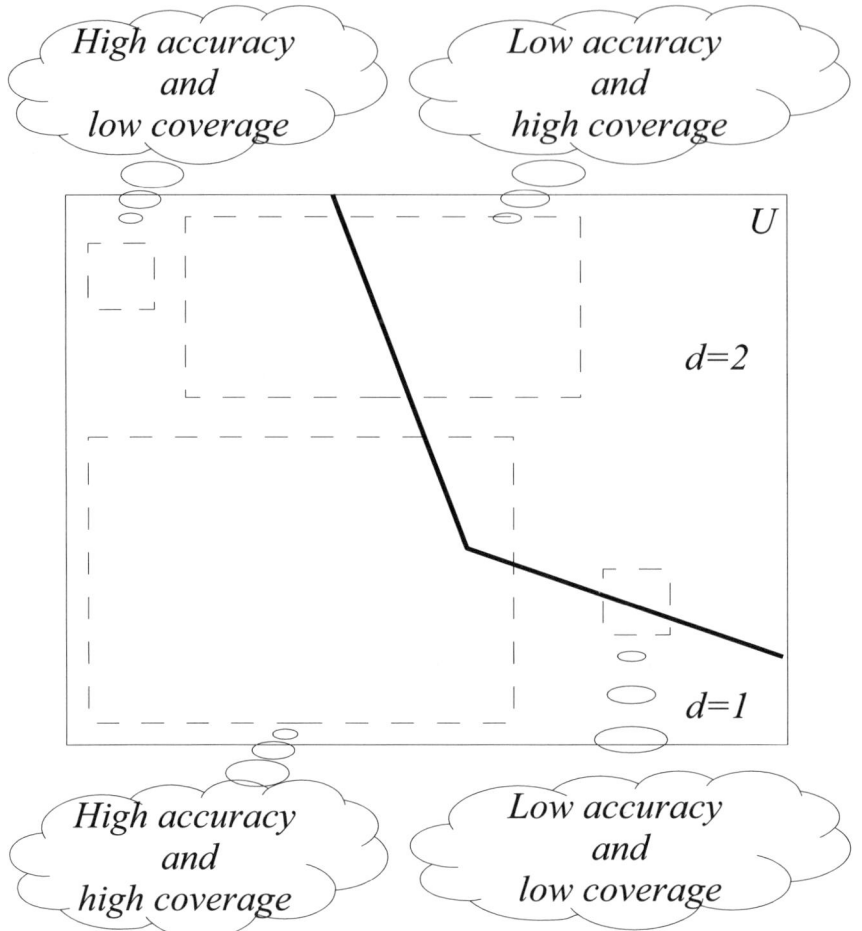

Fig. 4.2. Four Types of Rules

The Brazdil quality function q_B [17] is a product of accuracy and coverage.

$$q_B(Rule) = accuracy_{DT}(Rule) * e^{coverage_{DT}(Rule)-1}.$$

The Pearson quality function q_P [19] is based on the theory of contingency tables. Let

$$p_1 = card\left(\|\lambda \wedge \xi\|_{DT}\right) * card\left(\|\neg\lambda \wedge \neg\xi\|_{DT}\right),$$

$$p_2 = card\left(\|\lambda \wedge \neg\xi\|_{DT}\right) * card\left(\|\neg\lambda \wedge \xi\|_{DT}\right).$$

Using p_1 and p_2 we define

$$q_P(Rule) = \frac{(p_1 - p_2)^2}{card\left(\|\lambda\|_{DT}\right) * card\left(\|\xi\|_{DT}\right) * card\left(\|\neg\lambda\|_{DT}\right) * card\left(\|\neg\xi\|_{DT}\right)}.$$

An overview of rule quality formulas is given in [19].

It is easy to observe that the presented formulas yield values from the interval $[0, 1]$. One can also observe [19] that discussed quality functions are non-decreasing functions of accuracy and coverage.

Example 4.2. Let us consider data from Table 3.2. We compute different coefficients, for the rule

Rule : **if** $a_1 \in [0, 5]$ **and** $a_3 \in [0, 1]$ **then** $d = C$

obtained from the relative object related reduct $\{a_1, a_3\}$ for the objects x_2 and x_5.

For *Rule* we obtain the 2×2 contingency Table 4.3.

Table 4.3. Example of Contingency Table Representing the Quantitative Information about *Rule*

	$d = C$	$\neg (d = C)$	
$a_1 \in [0, 5] \wedge a_3 \in [0, 1]$	5	2	7
$\neg (a_1 \in [0, 5] \wedge a_3 \in [0, 1])$	0	2	2
	5	4	9

We compute support, accuracy and coverage:

$$support_{DT} (Rule) = card \left(\|a_1 \in [0, 5] \wedge a_3 \in [0, 1]\|_{DT} \cap \|d = C\|_{DT} \right) = 5,$$

$$accuracy_{DT} (Rule) = \frac{card \left(\|a_1 \in [0, 5] \wedge a_3 \in [0, 1]\|_{DT} \cap \|d = C\|_{DT} \right)}{card \left(\|a_1 \in [0, 5] \wedge a_3 \in [0, 1]\|_{DT} \right)} = \frac{5}{7},$$

$$coverage_{DT} (Rule) = \frac{card \left(\|a_1 \in [0, 5] \wedge a_3 \in [0, 1]\|_{DT} \cap \|d = C\|_{DT} \right)}{card \left(\|d = C\|_{DT} \right)} = 1.$$

We obtain the following qualities of *Rule* :

$$q_M (Rule) = w * accuracy_{DT}(Rule) + (1 - w) * coverage_{DT}(Rule) =$$

$$w * \frac{5}{7} + (1 - w) * 1 = 1 - \frac{2}{7} * w,$$

where $0 \leq w \leq 1$ is a parameter,

$$q_T (Rule) = \left(\frac{1}{2} + \frac{1}{4} accuracy_{DT}(Rule) \right) * accuracy_{DT}(Rule) +$$

$$(1 - \left(\frac{1}{2} + \frac{1}{4} accuracy_{DT}(Rule) \right)) * coverage_{DT}(Rule) =$$

$$\left(\frac{1}{2} + \frac{5}{28} \right) * \frac{5}{7} + \left(\frac{1}{2} - \frac{5}{28} \right) * 1 = 0.81,$$

$$q_B (Rule) = accuracy_{DT}(Rule) * e^{coverage_{DT}(Rule) - 1} = \frac{5}{7},$$

$$q_P (Rule) = \frac{(5 * 2 - 2 * 0)^2}{7 * 5 * 2 * 4} = 0.36.$$

4.4 Nearest Neighbor Algorithms

Learning in nearest neighbors algorithm consists of storing the presented data table. When a new object is encountered, a set of similar related objects is retrieved from memory and used to classify the new object.

The nearest neighbors method is an example of analogy-based reasoning. A reasoning system assumes that there is a data table providing the complete information about objects. When the system is asked about new object without decision it retrieves similar (analogous) objects from data table and the decision is completed on the basis of the information about the retrieved objects.

One advantage of nearest neighbors algorithm is that training is very fast. The second advantage is that the algorithm can learn complex relationships between condition and decision attributes.

One disadvantage of nearest neighbor approach is that the cost of classifying new objects can be high. This is due to the fact that nearly all computation takes place at classification time.

The second disadvantage is based on observation that nearest neighbor approach can be easily fooled by irrelevant attributes. Namely, the distance between objects is calculated based on all attributes of the data table. This lies in contrast to methods based on rough set approach that select only a subset of the attributes when forming a decision algorithm (set of rules). For example, consider applying nearest neighbors approach a problem in which each object is described by twelve attributes, but where only three of these attributes are relevant to determining the classification. In this case, objects that have identical values for the three relevant attributes may nevertheless be distant from one another in twelve dimensional space. As a result, the metric used by nearest neighbors algorithm - depending on all 12 attributes - will be misleading. The distance between neighbors will be dominated by the large number of irrelevant attributes.

One approach to overcoming this problem is to completely eliminate the least relevant attributes from the set of all attributes. The relevant attributes are extracted using rough set approach based, for example, on so called dynamic reducts. The details of the approach are discussed in Chapter 6. Next $k-NN$ method is applied to the relevant attributes only.

5 Selected Clustering Methods

In this chapter we recall the concept of clustering and discuss in detail some selected algorithms.

The organization of this chapter is as follows. In Section 5.1 selected clustering algorithms are discussed. In Section 5.2 the self-organizing system for information granulation is recalled. In Section 5.3 we discuss some clustering algorithms received by combining clustering with methods of the rough set theory. In Section 5.4 quality of information granulation is discussed.

5.1 From Data to Clusters

Clustering arises as an algorithmic framework for data mining. It is important task in several data mining applications including document retrieval and image segmentation. Let us observe that dividing n objects into nc clusters (granules) gives rise to a huge number of possible partitions, which is expressed in the form of the Stirling number of the second kind $S(n, nc)$. Stirling numbers of the second kind obey the recurrence relation $S(n, nc) = S(n-1, nc-1) + nc \cdot S(n-1, nc)$ with $S(n, 1) = 1$ and $S(n, n) = 1$. The Stirling numbers of the second kind are given by the explicit formula:

$$S(n, nc) = \sum_{j=1}^{nc} (-1)^{nc-j} \frac{j^{n-1}}{(j-1)!(nc-j)!} = \frac{1}{nc!} \sum_{j=1}^{nc} (-1)^{nc-j} \binom{nc}{j} j^n$$

In the case of $n = 107$ (see medical case study in Chapter 6) and $nc = 2, \ldots, 11$ the number of possible partitions is presented in Table 5.1.

This means that computation of optimal partition is a non-trivial task and it can not be solved by a simple increase of computational resources. Obviously, we need to resort to some optimization techniques the ones known as clustering methods.

There are many clustering methods, most of them designed for particular purposes and applications. They may be divided in the following categories: partitioning, model-based, hierarchical, density-based and grid-based. Partitioning algorithms perform clustering assigning labels to the individual object from

J. Stepaniuk: Rough - Gran. Comput. in Knowl. Dis. & Data Min., SCI 152, pp. 67–77, 2008.
springerlink.com

Table 5.1. Number of Possible Partitions for $n = 107$

nc	2	3	4	5	6	7	8	9	10	11
S(107,nc)	10^{32}	10^{50}	10^{63}	10^{73}	10^{80}	10^{87}	10^{92}	10^{96}	10^{100}	10^{104}

the dataset, hierarchical techniques create a structure explaining dependencies in the data, in density-based clustering density of data is examined and grid-based methods do clustering on the quantized space into a finite number of cells [59].

Partitioning algorithms are based on optimizing a criterion function that is the most common the sum-squared-error function. The most popular partitioning method is k-means (in the top 10 data mining algorithms [217]). The k-means algorithm is a simple iterative method to partition a given data set into a user specified number of clusters k. This algorithm has been discovered by several researchers across different disciplines (see e.g. [84] and references in [217]). The k-means algorithm can be placed in the larger context of hill-climbing algorithms. The algorithm works as follows: initially the data are grouped arbitrary, then in next steps the objects are moved to the nearest cluster to minimalize the criterion function. The clusters are determined by their centroids. Iterations are repeated until there is no objects to move. Although simple partitional algorithms suffer from some difficulties, among the other things, the number of clusters is required a priori and they find problems when clusters are large varied in size. The applied criterion function limits to processing only concentric data. Despite of the mentioned disadvantages the k-means algorithm is the most often used method in clustering.

Hierarchical clustering algorithms create a kind of tree of clusters, called dendrogram showing relationship among objects. Grouping into the desired number of clusters is achieved by cutting the dendrogram at an appropriate level. Dendrogram can be created in two way: agglomerative and divisive. Agglomerative approach starts from connected pairs of the most similar objects forming clusters. Then in followings steps the most similar clusters are joined. The number of clusters in every step is decreasing to one. Divisive approach works from top to down of the tree, starting from all objects forming one cluster. The clusters are split in two at every iteration until they are composed of one object. On account of method of distance calculation between clusters they can be divided in single-link [168], complete-link (hcl) [68] and minimum-variance [211] approaches.

In complete-link (or complete linkage) hierarchical clustering, we merge in each step the two clusters with the smallest maximum pairwise distance. In other words, in complete-link clustering (complete-linkage) clustering, the similarity of two link clustering clusters is the similarity of their most dissimilar members (see Figure 5.1). This results in a preference for compact clusters with small diameters over long, straggly clusters. This is a very popular technique offering good quality results.

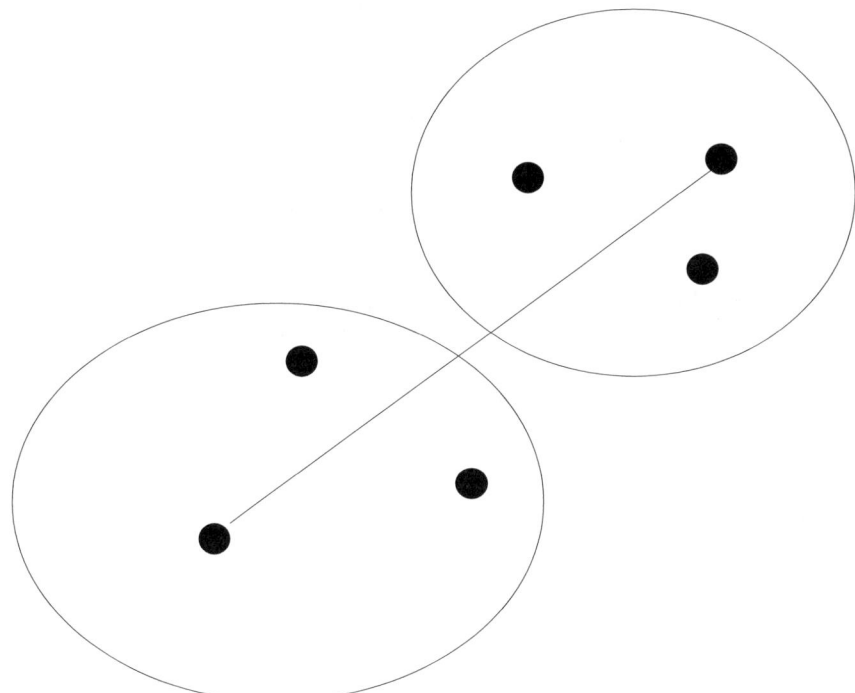

Fig. 5.1. Minimum Similarity used by the Hierarchical Complete-Link (hcl) Algorithm

In model-based clustering there is a hypothesized model for each of the clusters and the task is to find the best fitting of the parameters of that model [36]. Example model-based approach is neural networks with SOM [70]. Finite mixture distributions provide a flexible and mathematical-based approach to the clustering of data observed on random phenomena. We focus in the experiments with medical data on the use of normal mixture models, which can be used to cluster data. These mixture models can be fitted by maximum likelihood via the EM (ExpectationMaximization) algorithm (which is in the group of top ten data mining algorithms [217]) (for references see e.g. [215, 217]).

Density-based clustering technique groups neighboring objects basing on their density condition. This group of methods can cope with data of arbitrary shapes and is often used in spatial clustering. Typical density-based method is DB-SCAN (**D**ensity-**B**ased **S**patial **C**lustering of **A**pplications with **N**oise) [32]. We introduce the notion of the $\varepsilon-$ neighborhood. Given some object $x \in U$, the $\varepsilon-$neighborhood, denoted by $I_\varepsilon(x)$ is defined as:

$$I_\varepsilon(x) = \{y \in U : \delta(x, y) \leq \varepsilon\}$$

The form of the distance function δ implies the geometry of the $\varepsilon-$ neighborhood. Higher values of ε produce larger neighborhoods. We introduce another parameter, $minPts$, that tells us how many objects fall within a neighborhood.

Core object is an object with at least $minPts$ objects within a $\varepsilon-$ neighborhood.

We say that $x \in U$ is $y \in U$ directly density reachable with parameters ε and $minPts$ if and only if $x \in I_\varepsilon(y)$ and $card(I_\varepsilon(y)) \geq minPts$.

Objects $x \in U$ and $y \in U$ are density reachable if and only if there exists a chain of directly density reachable objects from x to y.

A cluster is defined as a maximal set of density connected objects (with respect to reachability).

The algorithm DBSCAN starts with an arbitrary object, and retrieves all object reachable from it with a distance no more than a given ε parameter. All such objects are therefore part of the same cluster, together with the arbitrary starting object, and are later used to recognize new nearby objects.

The DBSCAN algorithm consists of the following sequence of steps.

Set up the parameters of the neighborhood, ε and $minPts$:

1. Arbitrary select an object $x \in U$
2. Find (retrieve) all objects density reachable from x with respect to ε and $minPts$
3. If x is a core object, then the cluster is formed
4. Otherwise x is a border object, no objects are density-reachable from x and move on to the next object of U
5. Continue the process until all of the objects from U have been processed

Grid-based techniques quantize the space into a multi-resolution grid data structure and then do all operations on the quantized space. Time of processing data is independent directly on the data object and depends only on the number of cells in each dimension in quantized space. It makes the methods fast, but characterized by a small accuracy. Additionally, they have a tendency to identify clusters that do not exist. An example of grid-based algorithm is STING [210].

5.2   Self-Organizing Clustering System

The SOSIG (**S**elf-**O**rganizing **S**ystem for **I**nformation **G**ranulation) algorithm is a clustering system designed for detecting granules present in data. It is a successor of SArIS algorithm [213]. The algorithm SOSIG is described in detail in [75]. In this section we only recall main idea of the system.

SOSIG creates a network structure of connected objects forming clusters. Organization of the system, including as well the objects as the connections, is constructed on the basis of relationships between input data, without any external supervision. The structure points are representatives of input data, that is an individual object from the structure stands for one or more object from input set. In effect of this the number of representatives is much less than clustered data without lost of information.

Let us assume that input data is defined as an information system $IS = (U, A)$ [106], where $U = \{x_1, \ldots, x_n\}$ is a set of objects and $A = \{a_1, \ldots, a_k\}$ is a set of attributes. Result generated by SOSIG is also described by an information system $IS' = (Y, A \cup \{a_{gr}\})$, where the last attribute $a_{gr} : Y \rightarrow \{1, \ldots, nc\}$ denotes label of generated granule and $card(Y) \leq card(U)$ and $\forall x \in U \exists y \in Y (\delta(x, y) < NAT)$. In terminology of artificial immune systems, U can be interpreted as the set of antigens and Y can be interpreted as the set of antibodies (see [213]). The parameter NAT (network affinity threshold) defines neighborhood of objects from IS'. It directly influences level of granulation of the input set. Initial value of NAT is proportional to maximal number of nearest neighbor distances in the input set (see Equation 5.1).

$$NAT_{init} = \max(\{\min(\{\delta(x_i, x_j) : x_j \in U \ \& \ x_j \neq x_i\}) : x_i \in U\}) \qquad (5.1)$$

The following values of NAT are calculated from current state of the network (see Equation 5.2).

$$NAT = \frac{1}{rg} * \frac{\sum_{y_i \in Y} \min(\{\delta(y_i, y_j) : y_j \in Y \ \& \ y_j \neq y_i\})}{card(Y)} \qquad (5.2)$$

where $rg \in (0, \infty)$ is a resolution of granulation parameter. The most often value of granulation resolution parameter is 2.0 (estimated theoretically and confirmed in experiments) assigned to the most separated clusters. Decreasing the value of rg it is possible to identify granules in higher resolution. The NAT directly affects cluster formation as connections in network are determined if the objects are in their neighborhoods.

After initial phase, like normalization of data, calculation initial value of NAT, iterated steps of the algorithm follow. First, the system objects are assessed. The measure of their usefulness is a stimulation level s_l expressed by Equation 5.3.

$$s_l(y) = NAT - \min(\{\delta(y, x) : x \in U\}) \qquad (5.3)$$

Then useless objects are removed. It affects as well not stimulated objects as redundant ones. As redundant are determined points having the same input object in their neighborhood. The best of them stay in the network and also not redundant ones for other input data. This process controls the size of the network prevents forming excessively dense clusters. It results in compression phenomenon.

The remaining objects are re-connected and labeled. Components of the same granule have equal label, whereas the granule is determined by edges between the objects in the structure. Then there is calculated a new value of NAT parameter (see Equation 5.2). When stopping criterion is met, the algorithm is stopped after connections reconstruction. Otherwise following steps are carried out. As a stopping criterion there is considered a stable state of the network, that is the state of small fluctuations of network size and value of NAT. The last step is a procedure of replication of the all objects. Further modification of their attribute values allows searching better solution nearby examined network object. There is also a step introducing to the system object from input data not recognized yet. This operation avoids leaving not represented area in the training set.

Further classification of new as well as training objects can be performed using so-created structure. To assign a label to considered object it is necessary to determine neighborhood objects from network structure. The neighborhood of the object is defined by final value of the NAT (the last calculated value) of the SOSIG. The predominant value of the labels is given to the examined object.

5.3 Rough Clustering

In this section we recall some clustering methods combined with rough set methodology.

Clustering in relation to rough set theory is attracting increasing interest among researchers. In [93] rough sets are used to model the clusters in terms of lower approximation and boundary region (the part of an upper approximation that is not covered by a lower approximation). To initialize the rough k-means one has to select the weights of the lower approximation and the boundary region as well as the number of clusters. Mitra argued that a good initial setting of these parameters is one of the main challenges in rough set clustering. Genetic algorithms are used to tune the threshold, and relative importance of upper and lower approximation parameters of the sets. The Davies-Bouldin clustering validity index is used as the fitness function of the genetic algorithm, that is minimized. In [114] the performance of rough k-means algorithm [83] with respect to its compliance to the classical k-means was investigated. The following assumptions for rough k-means algorithm were made:

- An object belongs to one lower approximation at most.
- If an object is no member of any lower approximation it belongs to two or more upper approximations.
- A lower approximation is a subset of its corresponding upper approximation.

Let $(U, \{a_1, \ldots, a_p\})$ be an information system with $p > 0$ real valued attributes and $\{C_i \subseteq U : i = 1, \ldots, k\}$ be a set of clusters. We define m_i^{LOW} and m_i^{BN} by

$$m_i^{LOW} = \sum_{x \in LOW(C_i)} \frac{(a_1(x), \ldots, a_p(x))}{card(LOW(C_i))},$$

$$m_i^{BN} = \sum_{x \in BN(C_i)} \frac{(a_1(x), \ldots, a_p(x))}{card(BN(C_i))}.$$

The means m_i where $i = 1, \ldots, k$ are computed as weighted sums of the objects $x \in U$ in the lower approximation $LOW(C_i)$ (weight w_{LOW}) and the boundary region $BN(C_i))$ (weight w_{BN})

$$m_i = \begin{cases} w_{LOW} * m_i^{LOW} + w_{BN} * m_i^{BN} & \text{if } BN(C_i) \neq \emptyset \ \& \ LOW(C_i) \neq \emptyset \\ m_i^{LOW} & \text{if } BN(C_i) = \emptyset \ \& \ LOW(C_i) \neq \emptyset \quad (5.4) \\ m_i^{BN} & \text{if } BN(C_i) \neq \emptyset \ \& \ LOW(C_i) = \emptyset \end{cases}$$

The parameters $w_{LOW} \geq 0$ and $w_{BN} \geq 0$ correspond to the relative importance of the lower approximation and the boundary region, and $w_{LOW} + w_{BN} = 1$.
Then rough k-means clustering algorithm goes as follows:

1. Define the initial parameters: the weights w_{LOW} and w_{BN}, the number of clusters k and a threshold ε
2. Randomly assign the data objects to one lower approximation (and to the corresponding upper approximation).
3. Calculate the means according to Equation 5.4.
4. For each data object, determine its closest mean. If other means are not reasonably farer away as the closest mean (defined by the threshold ε) assign the data object to the upper approximations of these close clusters. Otherwise assign the data object to the lower approximation (and to the corresponding upper approximation) of the cluster of its closest mean.
5. Check convergence. If converged: Stop otherwise continue with Step 3.

If the upper approximation of each cluster is equal to its lower approximation, then the clusters are conventional clusters. Therefore, the boundary region $BN(C_i)$ (for $i = 1, \ldots, k$) is empty and Equation 5.4 reduce to conventional mean calculations.

In [92] the authors exploit the rough set based decision rules to obtain initial approximate mixture model parameters. Rough set theory offers a fast and robust solution to the initialization and local minima problems of iterative refinement clustering (like EM and k-means).

In [82] the time complexity of two approaches for obtaining intervals of clusters is discussed. Both the approaches use properties of rough sets. One approach is based on genetic algorithms, and the other is an adaptation of k-means algorithm.

Analysis and estimation of incomplete data is an increasingly important issue in many fields of knowledge discovery in databases. The paper [80] investigates missing data imputation techniques with the aim of constructing robust algorithms. Traditional clustering algorithms, e.g., k-means clustering, which are normally crisp, have been widely used in imputation. However, the crispness property makes the algorithms less practical, because an object could be assigned to more than one cluster. Integrating fuzzy sets into k-means clustering helps solve the crispness because the fuzzy membership function models the membership degree of an object in a cluster. Based on fuzzy set theory and rough set theory, the paper [80] presents three imputation algorithms, fuzzy k-means, rough k-means, and rough-fuzzy k-means.

In [48] an indiscernibility-based clustering method is presented. The main benefit of this method is that it can be applied to proximity measures that do not satisfy the triangular inequality. Additionally, it may be used with a proximity matrix thus it does not require direct access to the original attribute values.

In [1] was proposed the following approach based on clustering and rough sets:

1. Cluster formation via unsupervised clustering algorithms,
2. Database simplification and attribute selection via attribute discretization,
3. Decision rule extraction via rough set methods.

The approach was tested using biomedical databases.

In [126] the basic idea is to find reducts in an information system and apply them to any clustering procedure able to cope with discrete data. The authors apply the approach to a toy example of animal taxonomy in order to show its functionality.

In [85] the authors propose a rough–fuzzy c-means algorithm (RFCM), based on rough sets and fuzzy sets. While the membership function of fuzzy sets enables efficient handling of overlapping clusters, the concepts of lower and upper approximations deals with incompleteness in class definition. Each partition is represented by three parameters, namely, a cluster prototype, a crisp lower approximation, and a fuzzy boundary. Several quantitative measures based on rough sets evaluate the performance of the proposed algorithm. The authors observed that RFCM is superior to other c-means algorithms.

In [115] a rough k-medoids clustering algorithm was introduced. The algorithm was applied to four different data sets (synthetic, colon cancer, forest and control chart data). The authors compared the results of these experiments to rough k-means and discussed the strengths and weaknesses of the rough k-medoids.

5.4   Evaluation of Clustering

Together with specification of elementary granules it is necessary to define measures of granule quality [159]. In spite of the diversity of clustering algorithms, the aim of clustering techniques is detection of granules, that are possibly the most compact and separable. The compactness expresses how close the objects in a cluster are. For example, consider a variance of the objects. In this case the lower the value of the variance, the higher the compactness of the cluster. We are interested in compact clusters, thus low values of compactness are desirable. Separability expresses how distinct the clusters are. Some way of expressing separability is to compute inter cluster distances. We obtain most compact and separable granules with small values of intra cluster distances and large values of inter cluster distances.

To evaluate compactness and separation of discovered clusters there were proposed statistics so-called validity indices. Validity indices are designed to estimation of quality of obtained partitioning. Assessment the most optimal result needs calculation of validity indices for different values of algorithm's parameter, what usually is a number of clusters. The most commonly used are Dunn and Dunn-like statistics and Davies-Bouldin (DB) index [46]. Their advantage is exhibition no trends with respect to the number of clusters, therefore the minimum (DB) or maximum (Dunn) value indicate the most optimal partition. The Dunn's value for specified number of granules nc is defined by Equation 5.5. Let U be a set of objects and let C_i be a cluster, where $i = 1, \ldots, nc$. We assume that $nc > 1$.

$$D_{nc} = \min_{i=1,\ldots,nc} \left\{ \min_{j=i+1,\ldots,nc} \left(\frac{d(C_i, C_j)}{\max_{k=1,\ldots,nc} diam(C_k)} \right) \right\} \qquad (5.5)$$

where $d\left(C_i, C_j\right)$ is the dissimilarity function between two clusters C_i and C_j defined as

$$d(C_i, C_j) = \min_{x \in C_i, y \in C_j} d(x, y) \tag{5.6}$$

and $diam(C)$ is a diameter of a cluster defined as follows:

$$diam(C) = \max_{x, y \in C} d(x, y) \tag{5.7}$$

Follows the above definition the index value is large for compact clusters situated significantly far from one another.

The DB index is expressed by Equation 5.8. It is defined for the number of clusters equals nc.

$$DB_{nc} = \frac{1}{nc} \sum_{i=1}^{nc} \left(\max_{j=1,\ldots,nc, j \neq i} R_{ij} \right) \tag{5.8}$$

where

$$R_{ij} = \frac{stdev\left(C_i\right) + stdev\left(C_j\right)}{d(C_i, C_j)} \tag{5.9}$$

where $stdev\left(C_i\right)$ $\left(stdev\left(C_j\right)\right)$ denotes standard deviation of a cluster C_i $(C_j,$ respectively). The standard deviation of a cluster i is given by Equation 5.10.

$$stdev\left(C_i\right) = \frac{1}{|C_i|} \sqrt{\sum_{x \in C_i} \left(d\left(x, \bar{x}\right)\right)^2} \tag{5.10}$$

where $\bar{x}$ is a centroid of a cluster and $d\left(x, \bar{x}\right)$ is an Euclidean distance between the point x and the centroid $\bar{x}$.

The DB index measures the average similarity between each cluster and its most similar one, thus it is desirable to minimize this value.

Kaufman and Rousseeuw [74] proposed the Silhouette index (SI) to measure the strengths of clusters. For a given cluster $C \subseteq U$, this method assign to each object y_i of the cluster C a quantitative measure s_i, which indicates the membership of object in the cluster it has been assigned. Let a_i be the average distance of the object y_i to other objects in the cluster C and b_i is the average distance of y_i to objects in the nearest neighbor cluster besides its own. We define

$$s_i = \frac{b_i - a_i}{\max(a_i, b_i)}$$

This index s_i can take values from -1 to 1. When the index is zero, then the object y_i has equal distance to its cluster and its nearest neighbor cluster. If the index is positive, then the object y_i is closer to its cluster than other clusters.

If the index is negative, then the object y_i is wrongly assigned to the current cluster. The SI index is then defined in Equation 5.11.

$$SI = \frac{1}{card(U)} \sum_{y_i \in U} s_i \qquad (5.11)$$

Thus, if all objects are correctly assigned, then SI should be close to 1.

Although there are many proposed validity indices they suffer from limitations like assumption of distribution of clusters (the most often elliptical and spherical) or requirement of clusters of equal size and densities [46]. The last proposed index, CDbw is based on cluster compactness and separation. It is product of two components $Sep(nc)$ and $Intra\_dens(nc)$ (see Equation 5.12).

$$CDbw(nc) = Sep(nc) \cdot Intra\_dens(nc) \qquad (5.12)$$

where nc denotes a number of clusters in examined partitioning. The value $Intra\_dens(nc)$ depends on density of granules (described in the following text) and $Sep(nc)$ is a measure of separability of clusters defined as follows:

$$Sep(nc) = \frac{\sum_{i=1}^{nc} \sum_{j=1,i \neq j}^{nc} \min d(clos\_rep_i, clos\_rep_j)}{1 + Inter\_dens(nc)} \qquad (5.13)$$

Separability is proportional to the sum of distances between the closest representative points $close\_rep$ from pair-wise clusters and inversely proportional to measure of density between clusters $Inter\_dens(nc)$. Representative points are objects from a training set selected by Farthest First (FF) algorithm for every cluster. The FF algorithm works as follows: initially the cluster centroid is determined. Then there is selected the first representative point from the training data located the farthest from the centroid. In the following steps are selected the farthest objects from the previously determined representatives. The steps are repeated until the required number r of representatives is reached. Density between clusters is expressed by Equation 5.14. It is desirable to generate partitioning of the lowest inter-cluster density.

$$Inter\_dens(nc) = \sum_{i=1}^{nc} \sum_{j=1,j \neq i}^{nc} \left(\frac{d(clos\_rep_i, clos\_rep_j)}{stdev(C_i) + stdev(C_j)} \cdot density(u_{ij}) \right) \qquad (5.14)$$

The component $stdev(C)$ is the standard deviation of clusters previously defined by Equation 5.10 and $density(u_{ij})$ density of input objects around the point u_{ij} defined by Equation 5.15. The point u_{ij} is the middle point of the line segment defined by the closest clusters' representatives $close\_rep_i$ and $close\_rep_j$.

$$density(u_{ij}) = \frac{\sum_{x \in C_i \cup C_j} f(x, u_{ij})}{card(C_i) + card(C_j)} \qquad (5.15)$$

The $density(u_{ij})$ represents the percentage of points in the cluster i and the cluster j that belong to the neighborhood of u_{ij}. This neighborhood is defined to be a hyper-sphere with center u_{ij} and radius the average standard deviation of the clusters between which the density is estimated. The function $f(x, u_{ij})$ is defined as:

$$f(x, u_{ij}) = \begin{cases} 0 \text{ if } d(x, u_{ij}) > stdev\,(C_i) + stdev\,(C_j))/2 \\ 1 \text{ otherwise.} \end{cases} \quad (5.16)$$

The second component of Equation 5.12 determines the average density within clusters and is defined as the percentage of points that belong to the neighborhood of representative points of the considered clusters. The goal is the density within clusters to be significant high. $Intra\_dens(nc)$ is given by the following equation:

$$Intra\_dens(nc) = \frac{1}{nc} \sum_{i=1}^{nc} \frac{1}{r} \sum_{v_{ij} \in C_i} \frac{density(v_{ij})}{stdev\,(C_i)} \quad (5.17)$$

where

$$density(v_{ij}) = \sum_{x \in C_i} g(x, v_{ij}) \quad (5.18)$$

The function $g(x, v_{ij})$ is described by Equation 5.19.

$$g(x, v_{ij}) = \begin{cases} 0 \text{ if } d(x, v_{ij}) > stdev\,(C_i) \\ 1 \text{ otherwise.} \end{cases} \quad (5.19)$$

To determine a good clustering scheme it is required to find a maximum value of CDbw.

In general, while the presented validity indexes are useful, in many cases they may produce inconclusive results. They offer only some guidelines and do not decisively point at the unique number of granules.

6 A Medical Case Study

Many interesting applications of rough set methods are reported. Let us mention only some of medical applications: risk pattern identification in the treatment of infants with respiratory failure [6], treatment of duodenal ulcer by HSV [111], analysis of data from peritoneal lavage in acute pancreatitis [165], knowledge acquisition in nursing [45], medical databases (e.g. headache, meningitis, CVD) analysis ([204, 205]), image analysis for medical applications ([61, 94]), surgical wound infection [66], preterm birth prediction [45], medical decision–making on board space station Freedom (NASA Johnson Space Center) [45], diagnosing in progressive encephalopathy [209], data preparation for data mining in medical data sets [56], selection of important attributes for medical diagnosis systems [57], visualization of rough set decision rules for medical diagnosis systems [58], automatic detection of speech disorders [24], rough set-based filtration of sound applicable to hearing prostheses [23], discovery of attribute dependences in diabetes data [187].

In this chapter the applications of the rough set theory to identify the most relevant attributes and to induce decision rules from a real life medical data set are discussed (see also [187, 189]). Information granulation using clustering is also investigated [197]. The real life medical data set concerns children with diabetes mellitus. Three methods are considered for identification of the most relevant attributes. The first method is based on the notion of reduct and its stability. The second method is based on particular attribute significance measured by relative decrease of positive region after its removal. The third method is inspired by the wrapper approach, where the classification accuracy is used for ranking attributes. The rough set approach additionally offers the set of decision rules. For the rough set based reduced data application of nearest neighbor algorithms is also investigated. The presented methods are general and one can apply all of them to different kinds of data sets.

The structure of the chapter is as follows. The description of clinical data is presented in Section 6.1. The searching for optimal attribute subsets is discussed in Section 6.2. In Section 6.3 application of nearest neighbors algorithms is investigated. In Section 6.4 discovery of decision rules is investigated. In Section 6.5

J. Stepaniuk: Rough - Gran. Comput. in Knowl. Dis. & Data Min., SCI 152, pp. 79–96, 2008.
springerlink.com

experiments with tolerance thresholds are presented. In Section 6.6 experiments with clustering algorithms are presented.

6.1 Description of the Clinical Data

There are two main forms of diabetes mellitus:

- type 1(insulin-dependent),
- type 2 (non-insulin-dependent).

Type 1 usually occurs before age 30, although it may strike at any age. The person with this type is usually thin and needs insulin injections to live and dietary modifications to control his or her blood sugar level. Type 2 usually occurs in obese adults over age 40. It's most often treated with diet and exercise (possibly in combination with drugs that lower the blood sugar level), although treatment sometimes includes insulin therapy.

We consider data about children with insulin-dependent diabetes mellitus (type 1). Insulin-dependent diabetes mellitus is a chronic disease of the body's metabolism characterized by an inability to produce enough insulin to process carbohydrates, fat, and protein efficiently. Treatment requires injections of insulin.

Complications may happen when a person has diabetes. Some effects, such as hypoglycemia, can happen any time. Others develop when a person has had diabetes for a long time. These include damage to the retina of the eye (retinopathy), the blood vessels (angiopathy), the nervous system (neuropathy), and the kidneys (nephropathy). The typical form of diabetic nephropathy has large amounts of urine protein, hypertension, and is slowly progressive. It usually doesn't occur until after many years of diabetes, and can be delayed by tight control of the blood sugar. Usually the best lab test for early detection of diabetic nephropathy is measurement of microalbumin in the urine. If there is persistent microalbumin over several repeated tests at different times, the risk of diabetic nephropathy is higher. Normal albumin excretion is less than 20 microgram/min (less than 30 mg/day). Microalbuminuria is 20-200 microgram/min (30-300 mg/day).

Twelve condition attributes, which include the results of physical and laboratory examinations and one decision attribute (microalbuminuria) describe the database used in our experiments. The database is shown at the end of the paper [189]. The data collection so far consists of 107 cases. The collection is growing continuously as more and more cases are analyzed and recorded. Out of twelve condition attributes eight attributes describe the results of physical examinations, one attribute describes insulin therapy type and three attributes describe the results of laboratory examinations. The former eight attributes include sex, the age at which the disease was diagnosed and other diabetological findings. The latter three attributes include the criteria of the metabolic balance, hypercholesterolemia and hypertriglyceridemia. The decision attribute describes the presence or absence of microalbuminuria. All this information is collected during treatment of diabetes mellitus.

Table 6.1. Names and Types of Attributes

Symbol	Attribute name	Attribute type
a_1	Sex	string
a_2	Age of disease diagnosis (years)	integer
a_3	Disease duration (years)	integer
a_4	Appearance diabetes in the family	string
a_5	Insulin therapy type	string
a_6	Respiratory system infections	string
a_7	Remission	string
a_8	HbA1c	float
a_9	Hypertension	string
a_{10}	Body mass	string
a_{11}	Hypercholesterolemia	string
a_{12}	Hypertriglyceridemia	string
d	Microalbuminuria	string

Table 6.2. Ranges, Means and Standard Deviations of the Numerical Attributes

Class	Attribute	Range	Mean	Std. dev.
Yes	a_2	$[1, 17]$	9.75	4.01
&	a_3	$[2, 13]$	5.86	2.02
No	a_8	$[4.3, 12.73]$	8.31	1.48
	a_2	$[2, 16]$	10.82	3.02
Yes	a_3	$[2, 13]$	5.93	1.98
	a_8	$[4.3, 11.61]$	8.46	1.56
	a_2	$[1, 17]$	8.57	4.63
No	a_3	$[3, 13]$	5.78	2.08
	a_8	$[5.46, 12.73]$	8.14	1.39

Table 6.3. Attributes and Their Values after Discretization

Attribute	Attribute values
a_1	f, m
a_2	$< 7, [7, 13), [13, 16), \geq 16$
a_3	$< 6, [6, 11), \geq 11$
a_4	yes, no
a_5	KIT, KIT_IIT
a_6	yes, no
a_7	yes, no
a_8	$< 8, [8, 10), \geq 10$
a_9	yes, no
a_{10}	<3, 3-97, >97
a_{11}	yes, no
a_{12}	yes, no
d	yes, no

Table 6.4. Characterization of Patients Group after Discretization

	%	Count
Total number of patients	100	107
Sex		
Male	54.21	58
Female	45.79	49
Age of disease diagnosis (years)		
< 7	22.43	24
[7, 13)	49.53	53
[13, 16)	22.43	24
≥ 16	5.61	6
Disease duration (years)		
< 6	51.40	55
[6, 11)	42.99	46
≥ 11	5.61	6
HbA1c		
< 8	42.99	46
[8, 10)	42.06	45
≥ 10	14.95	16
Microalbuminuria		
yes	52.34	56
no	47.66	51

Attribute names and types can be found in Table 6.1.

The range, the mean and the standard deviation of the numerical attributes can be found in Table 6.2. Additionally attributes with numeric values were discretized. Although several algorithms for automatic discretization exist (for overviews see [95]), in this analysis discretization was done manually according to medical norms. Attributes and their values after discretization are presented in Table 6.3. Basic data information after discretization is presented in Table 6.4.

6.2   Relevance of Attributes

One can measure the importance of attributes with respect to different aspects. One can also consider different strategies searching for the most important subset of attributes. For example one can exhaust all possible subsets of the set of condition attributes and find the optimal ones. In general, its complexity (the number of subsets need to be generated) is $O\left(2^{card(A)}\right)$, where $card\left(A\right)$ is a number of attributes. This strategy is very time consuming. Therefore we consider less time consuming strategies.

In this section the relevance of attributes is evaluated and compared using three methods.

6.2.1   Reducts Application

We compute the accuracy of approximation of decision classes. From Table 6.5 one can observe that both decision classes are definable by twelve condition attributes.

Table 6.5. Accuracy of Approximation of Decision Classes

Decision class	Yes	No
Number of patients	56	51
Cardinality of lower approximation	56	51
Cardinality of upper approximation	56	51
Accuracy of approximation (α)	1.0	1.0

There are six reducts. Three reducts with nine attributes and three reducts with ten attributes. Reducts are presented in Table 6.6. Sign "$+$" means occurrence of the attribute in a reduct. Stability of reducts was verified on subtables. This idea was inspired by the concept of dynamic reducts [5]. Based on experimental verification, reducts for full data table are more stable than other attribute subsets. For example in one experiment we choose 30 subtables starting from 90% to 99% of all objects in data table, thus we consider 300 subtables. Six mentioned above reducts were also reducts at least in 69% from 300 subtables and other subsets were reducts in less than 10% of subtables.

In the Table 6.6 stability of the reducts based on four experiments is also presented. We consider 300 subtables in every experiment. The sampling strategy is the following: subtables are sampled on 10 equally spaced levels with 30 samples per level. In the following four experiments we consider different sampling levels:

- Experiment 1: 60%, 64%, ..., 96% of the original table.
- Experiment 2: 70%, 73%, ..., 97% of the original table.

Table 6.6. Reducts, Their Stability and Classification Accuracy

Attribute/Reduct	B_1	B_2	B_3	B_4	B_5	B_6
a_1	+	+	+	+	+	+
a_2	+	+	+	+	+	+
a_3	+	+	+	+	+	+
a_4	-	+	+	+	+	+
a_5	+	+	+	+	+	+
a_6	+	-	-	-	+	+
a_7	-	+	+	+	-	-
a_8	+	+	+	+	+	+
a_9	+	-	-	+	-	+
a_{10}	+	-	+	-	+	+
a_{11}	+	+	-	-	+	-
a_{12}	+	+	+	+	+	+
Stability of the reduct	65%	59%	58%	54%	47%	43%
Classification accuracy	63%	77%	71%	76%	68%	70%

- Experiment 3: 80%, 82%, ..., 98% of the original table.
- Experiment 4: 90%, 91%, ..., 99% of the original table.

In all experiments we consider subtables with at least 60% of the original table to preserve representability. On the other hand from evaluations presented in [5] we deduce that the number of at least 300 subtables is enough for good estimation of the stability coefficient.

For every reduct one can also compute classification accuracy based on leave-one-out method. The results are presented in the last row of the Table 6.6.

From the above analysis we infer that the reduct B_2 is a relatively stable subset of attributes with high classification accuracy of generated rules.

6.2.2 Significance of Attributes

For the set of all condition attributes the quality of approximation of classification is equal to 1.

In the first step we consider attributes that are in all six reducts. Thus we consider attributes in core. The quality of approximation is equal to 0.76. The ranking of core attributes is presented in Table 6.7. The idea is to evaluate each individual attribute with the significance measure. This evaluation results in a value attached to an attribute. Attributes are then sorted according to the values. The attribute with the least significance (the smallest contribution to the quality of the approximation) is removed and the process is repeated. One can stop the algorithm when the quality of approximation equals zero.

Table 6.7. Ranking Core Attributes

Attribute removed	Resulting quality of approximation
a_{12}	0.74
a_5	0.55
a_1	0.40
a_3/a_8	0.09
a_2	0

6.2.3 Wrapper Approach

We consider method inspired by wrapper approach [69]. The subsets of attributes are evaluated based on the cross-validation result.

We recall that the cross-validation is based on the following procedure. Choose $k > 1$ and partition the available data table $DT = (U, A \cup \{d\})$ into disjoint data subtables $DT_i = (U_i, A \cup \{d\})$ of equal size, where $i = 1, \ldots, k$. For i from 1 to k use DT_i for the test set, and the remaining data for training set. Compute classification accuracy for all k experiments. The average classification accuracy is a result of cross-validation test. Leave-one-out is a special case of cross-validation procedure. In leave-one-out cross-validation the set of $card(U) = n$ objects is repeatedly divided into a training set of size $n - 1$ and test set of size 1, in all possible ways.

In the succeeding steps of the analysis that attribute is removed which removal leads to the best result of the cross-validation test. Let $DT = (U, A \cup \{d\})$ be a data table. The general scheme of the algorithm is as follows:

$B := A$;
Repeat $B := B - \{a\}$, where $a = \arg\max_{a \in B} \left\{ AC \left(DT_{B-\{a\}} \right) \right\}$.
Until Stop_Condition;

where $DT_{B-\{a\}} = (U, (B - \{a\}) \cup \{d\})$ and the resulting accuracy coefficient is $AC \left(DT_{B-\{a\}} \right)$.

The partial results of the analysis are presented in Table 6.8. The leave-one-out test was used for accuracy estimation. The best result 79.44% was obtained for six attributes. The further removal of attributes thus not led to the increase of classification accuracy.

Table 6.8. Classification Accuracy and Quality of Approximation (γ)

Attribute	I	II	III	IV	V	VI
a_1	63.55	62.62	64.49	62.62	62.62	60.75
a_2	67.29	64.49	68.22	66.36	71.96	65.42
a_3	69.16	66.36	67.29	66.36	66.36	69.16
a_4	64.49	63.55	71.03	71.03	71.96	69.16
a_5	69.16	69.16	68.22	71.96	72.90	**79.44**
a_6	70.09	**72.90**	-	-	-	-
a_7	69.16	68.22	71.03	72.90	71.03	74.77
a_8	62.62	62.62	64.49	65.42	66.36	62.62
a_9	**71.03**	-	-	-	-	-
a_{10}	68.22	70.09	**76.64**	-	-	-
a_{11}	68.22	70.09	71.03	73.83	**73.83**	-
a_{12}	68.22	70.09	73.83	**74.77**	-	-
γ	1.0	1.0	1.0	0.98	0.95	0.80

Every method allows to analyze data from different angle. Combining the results of the three methods one can find the following three condition attributes as the most relevant: *Age of disease diagnosis*, *HbA1c* and *Disease duration*. This result is consistent with the general medical knowledge about this disease.

6.3   Rough Set Approach as Preprocessing for Nearest Neighbors Algorithms

In this section we discuss the experiments with nearest neighbor algorithms (see for example [69, 89], for more details). The nearest neighbor algorithm retains the entire training data set during learning. This algorithm assumes all objects correspond to points in n-dimensional space. In our experiments the nearest neighbors of an object are defined in terms of the Euclidean distance.

More precisely, let $DT = (U, A \cup \{d\})$ be a decision table, for every two objects $x, y \in U$ the Euclidean distance is defined by

$$E(x, y) = \sqrt{\sum_{a \in A} (a(x) - a(y))^2} \qquad (6.1)$$

A more general definition calculates the distance as

$$distance(x, y) = \sqrt{\sum_{a \in A} w_a * difference(a(x), a(y))^2} \qquad (6.2)$$

where w_a is a non-negative weight value assigned to attribute a and the difference between attribute values is defined by Equation 2.6.

Nearest neighbor algorithms are especially susceptible to the inclusion of irrelevant attributes in the data set, and several studies has shown that the classification accuracy degrades as the number of irrelevant attributes is increased (see e.g. [76]). Therefore we use in our experiments relevant subsets of attributes (based on rough set analysis).

Before applying nearest neighbors method to diabetes data, the values of non-numerical attributes were converted into numerical data in the manner described in Table 6.9.

Table 6.9. Conversion of Non-Numeric Attributes

	Attribute values
a_1	f - 0, m - 1
a_4	yes - 1, no - 0
a_5	KIT - 1, KIT_IIT - 0
a_6	yes - 1, no - 0
a_7	yes - 1, no - 0
a_9	yes - 1, no - 0
a_{10}	<3 - 0, 3-97 - 1, >97 - 2
a_{11}	yes - 1, no - 0
a_{12}	yes - 1, no - 0
d	yes - 1, no - 0

For number $k \in \{1, \ldots, 10\}$ of nearest neighbors and different attribute subsets (three most important attributes, all attributes and six reducts) we obtain the leave-one-out results presented in Table 6.10. The best results are obtained for the set $A_3 = \{a_2, a_3, a_8\}$.

We also compare the obtained results with linear discriminant classifier (for subsets of attributes as in nearest neighbors method). The idea of this procedure is to divide object set by a series of lines in two dimensions, planes in three dimensions and, generally hyperplanes in many dimensions (see for example [89], for more details). New objects are classified according to the side of the hyperplane that they fall on.

Table 6.10. Nearest Neighbors Method

k	A_3	A	B_1	B_2	B_3	B_4	B_5	B_6
1	73.83	73.83	71.96	76.64	70.09	70.09	81.31	73.83
2	90.65	85.98	86.92	87.85	87.85	86.92	87.85	86.92
3	75.70	72.90	72.90	75.70	71.96	73.83	73.83	71.96
4	83.18	79.44	79.44	82.24	80.37	80.37	80.37	78.50
5	79.44	72.90	71.96	72.90	73.83	75.70	73.83	73.83
6	85.98	83.18	82.24	84.11	85.05	85.05	83.18	83.18
7	79.44	76.64	76.64	76.64	76.64	80.37	75.70	78.50
8	82.24	82.24	80.37	82.24	84.11	83.18	80.37	82.24
9	78.50	77.57	77.57	74.77	80.37	80.37	77.57	77.57
10	81.31	82.24	82.24	82.24	82.24	82.24	81.31	82.24

The method requires numerical attribute vector, therefore we use conversion of the values of non-numerical attributes into numerical data (see Table 6.9) before applying linear discriminants.

In Table 6.11 we present the results obtained by linear discriminant classifier (and attribute subsets as for nearest neighbors method) with leave-one-out test.

Table 6.11. Linear Discrimination

A_3	A	B_1	B_2	B_3	B_4	B_5	B_6
72.53	67.17	66.37	65.21	58.88	60.66	66.37	66.37

The best result is also in the case of the set A_3 with three attributes, but is about 18% worse than for two nearest neighbors method.

6.4 Discovery of Decision Rules

Further analysis of the data table consists in determining the relationships between values of the attributes and presence or absence of microalbuminuria i.e. looking for representation of these relationships in the form of decision rules.

The exhaustive generation of object related reducts and rules from reducts was used to perform the experiment.

In the current experiment the data set has been split three times and all parts of the experiment have been duplicated for each of the three splits. This approach lessens the impact of randomness in the results.

The following procedure was repeated for $i = 1, 2, 3$. The data table $DT = (U, A \cup \{d\})$, where $card(U) = 107$ and $A = \{a_1, \ldots, a_{12}\}$ was split randomly into a testing set $DT_{test_i} = (U_{test_i}, A \cup \{d\})$ and a training set $DT_{train_i} = (U_{train_i}, A \cup \{d\})$ containing approximately 33% ($card(U_{test_i}) = 35$) and 67% ($card(U_{train_i}) = 72$) of the objects in DT, respectively.

We used Michalski's quality function (with weight w of accuracy equal to 0) discussed in Section 4.3 for rule filtering. Thus, this experiment corresponds to filtering according to coverage. For summary of obtained results see Table 6.12.

Table 6.12. Quality and Number of Rules

Quality	Number of rules			
	DT	DT_{train_1}	DT_{train_2}	DT_{train_3}
$(0, 0.05)$	472	223	291	304
$[0.05, 0.1)$	204	189	262	220
$[0.1, 0.15)$	43	88	73	51
$[0.15, 0.2)$	16	32	19	17
$[0.2, 0.25)$	2	13	2	7
≥ 0.25	1	9	4	1
> 0	738	554	651	600

The difference between classification accuracy obtained for unfiltered rule set (quality threshold equal to zero) and the rules with quality greater or equal than a given threshold is presented in Table 6.13.

Table 6.13. Difference in Classification Accuracy

Quality threshold	DT_{test_1}	DT_{test_2}	DT_{test_3}
0.00	0%	0%	0%
0.05	3%	3%	3%
0.10	0%	3%	0%
0.15	-3%	0%	-20%
0.20	-11%	-29%	-34%
0.25	-20%	-46%	-57%

We present examples of rules selected by medical experts from the set of all rules with $q_M \geq 0.1$, where rules are generated from full data table DT.

if $a_2 \in [7, 13)$ and $a_3 \in [6, 11)$ and $a_{11} = yes$ then $d = yes$
$accuracy_{DT} = 1$, $coverage_{DT} = 0.25$
if $a_1 = m$ and $a_2 < 7$ and $a_4 = KIT$ then $d = no$
$accuracy_{DT} = 1$, $coverage_{DT} = 0.20$.
if $a_1 = m$ and $a_3 < 6$ and $a_4 = KIT$ and $a_8 < 8$ and $a_9 = no$ then $d = no$
$accuracy_{DT} = 1$, $coverage_{DT} = 0.20$.
if $a_2 < 7$ and $a_3 \in [6, 11)$ then $d = no$
$accuracy_{DT} = 1$, $coverage_{DT} = 0.18$.
if $a_1 = f$ and $a_3 \in [6, 11)$ and $a_{11} = yes$ then $d = yes$
$accuracy_{DT} = 1$, $coverage_{DT} = 0.14$.

6.5 Experiments with Tolerance Thresholds

In this section we discuss selected results of numerous experiments performed by the author with uncertainty functions defined by distance measures and threshold functions.

We present results for three data tables $(U, A \cup \{d\})$, $(U, A_3 \cup \{d\})$ and $(U, A - A_3 \cup \{d\})$ with twelve, three $A_3 = \{a_2, a_3, a_8\}$ and nine condition attributes, respectively. We assume that for all non-numerical attributes the uncertainty function is defined in the standard way, i.e., for all $a \in A - A_3$ $I_a(x) = \{y \in U : a(x) = a(y)\}$.

The following thresholds have been found for attributes from A_3 : $\varepsilon_{a_2} = 0.063$, $\varepsilon_{a_3} = 0.182$, $\varepsilon_{a_8} = 0.16$ in the optimization process with respect to classification accuracy.

We compute the accuracy of approximation of decision classes. From Table 6.14 one can observe that both decision classes are only roughly definable by condition attributes with above thresholds.

We compare the results of leave-one-out test for attribute subsets A_3, A and $A - A_3$. The best results are presented in Table 6.15.

Table 6.14. Accuracy of Approximation of Decision Classes with Tolerance Thresholds

Attribute set	Decision class	Yes	No
	Cardinality of lower approximation	53	46
A	Cardinality of upper approximation	61	54
	Accuracy of approximation (α)	0.87	0.85
	Cardinality of lower approximation	11	13
A_3	Cardinality of upper approximation	94	96
	Accuracy of approximation (α)	0.12	0.14
	Cardinality of lower approximation	42	34
$A - A_3$	Cardinality of upper approximation	73	65
	Accuracy of approximation (α)	0.58	0.52

Table 6.15. Leave-One-Out Results

Attribute subset	Classification accuracy
A_3	79.44
A	66.36
$A - A_3$	63.55

Comparing results from Table 6.15 we obtain one more argument that attributes a_2, a_3 and a_8 are very relevant for prediction of d.

Discussing obtained decision rules with medical doctors, we observe that sometimes the obtained rules are not very informative for medical experts. More precisely the selectors in obtained rules are not necessarily in exact correspondence with norms used by medical doctors.

This observation can be extended to two different knowledge discovery problems, namely, prediction and description. Experiments with tolerance thresholds and also in some sense with the nearest neighbors method are showing that on the one hand the discretization of numerical attributes based on medical norms can be not optimal with respect to classification accuracy. On the other hand the understanding of decision rules by medical doctors is better when the discretization is based on medical norms, than in case of decision rules with selectors based on tolerance thresholds.

6.6 Experiments with Clustering Algorithms

The aim of the experiments carried out is to present information granules obtained in clustering process. To create the granules there were selected algorithms k-means, hierarchical complete-link (hcl), EM, DBSCAN and SOSIG (see Chapter 5).

Some preliminary results of granulation of not normalized data has been presented in [197].

Normalization consists of transforming numerical values into a specific range. The attributes have been normalized to interval $[0, 1]$. We normalize each numerical attribute a using (for any $x \in U$) the transformation

$$a_{new}(x) = \frac{a(x) - \min_a}{\max_a - \min_a}.$$

Table 6.16. Evaluation of Clusterings of the Diabetes Data Including the Decision Attribute Created by Algorithms: k-means and EM

Method	index	optimal value	nc
k-means	Dunn	0.43	6-9
	DB	1.51	10
	SI	0.18	10
	CDbw	1.82	3
EM	Dunn	0.43	9,10
	DB	1.79	10
	SI	0.15	2
	CDbw	1.85	3

Table 6.17. Evaluation of Clusterings of the Diabetes Data Excluding the Decision Attribute Created by Algorithms: k-means and EM

Method	index	optimal value	nc
k-means	Dunn	0.48	10
	DB	1.35	9
	SI	0.19	9,10
	CDbw	2.18	2
EM	Dunn	0.48	9,10
	DB	1.38	4
	SI	0.16	2
	CDbw	1.16	2

The clustering was executed with the decision attribute (microalbuminuria) present and without it. The algorithms k-means and EM have been run for $nc = 2, 3, ..., 10$, the method SOSIG for various values of granulation resolution rg and DBSCAN for various values of parameter ε with $minPts = 3$ (the ε parameter stands for radius of density neighborhood and $minPts$ denotes minimal number of objects in single cluster). Every partitioning is assessed by indices Dunn's, DB, SI and CDbw. The results containing less than 90% objects from input set clustered were not taken into consideration.

Table 6.16 presents evaluation of granulation of the 13 dimensional diabetes data. Table 6.17 presents evaluation of granulation of the same set excluding the decision attribute microalbuminuria. There is high discrepancy in optimal partitioning indicated by the values in Tables 6.16 and 6.17. The most often distinguished is clustering containing 8, 9 and 10 clusters, however there are also indicated results containing 2, 3 and 4 groups.

In DBSCAN result for $\varepsilon < 1.1$ regardless of $minPts$ value only 20-30 objects were clustered, whereas for $\varepsilon > 1.1$ all input objects were assigned to one cluster.

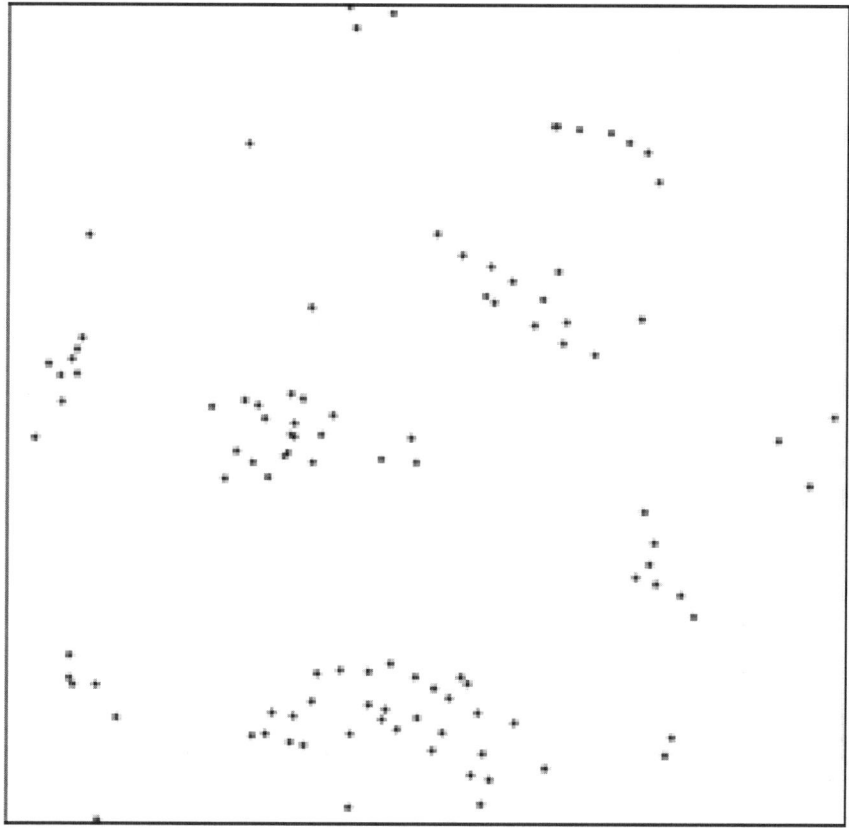

Fig. 6.1. MDS Plot of Information System with Attributes $a_2, a_3, a_8, a_9, a_{11}, a_{12}, d$

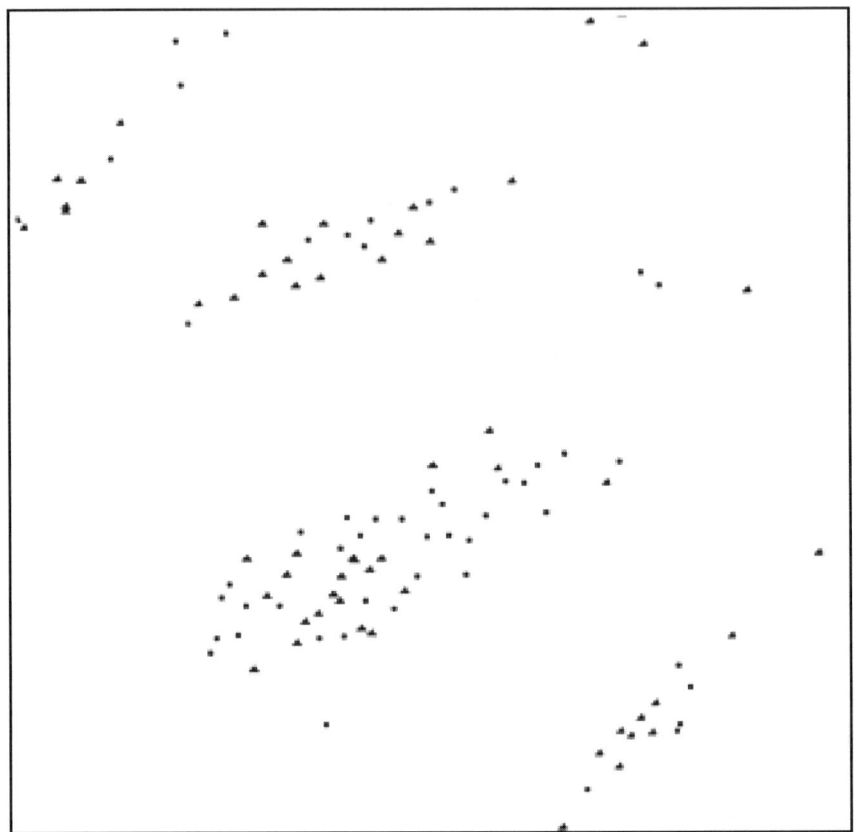

Fig. 6.2. MDS Plot of Information System with Attributes $a_2, a_3, a_8, a_9, a_{11}, a_{12}$

The only partitioning is obtained for $\varepsilon = 1.1$ containing 2 groups of 99 and 8 elements. Calculated assessment indices were DB=2.24, Dunn=0.36, CDbw=1.90, SI=0.12. Since high difference in sizes of clusters and low quality indicated the result is not taken in further considerations.

Similar results were obtained in SOSIG granulation. There was also threshold value of rg parameter ($rg = 2.2$ in 13 dimensional set and $rg = 2.1$ in 12 dimensional set grouping) below which not enough input objects were clustered and the result contained representatives not connected to each other. Over the threshold value the partitioning was composed of all objects forming one cluster.

Considering described before examples and ambiguous results of granulation 13 and 12 attributes of diabetes data one can assume there is absence of compact and separable clusters in the data. Presence of microalbuminuria attribute exerts an influence on obtained clustering, however separability of groups is low in both cases. This is concluded on the basis of highly diverse assessment values for parametrical clustering methods k-means, hierarchical and EM. More effortless and unambiguous interpretation is in case of SOSIG result. There is or not enough number of input objects clustered (the value of rg parameter is too low) or all the objects are forming one cluster. Formation one group implies the objects are equally similar to one another.

In further exploration of the data set, six attributes were chosen based on medical experts suggestion. It was selected set $\{a_2, a_3, a_8, a_9, a_{11}, a_{12}\}$. The multi-dimensional scaling (MDS) plot [22] of seven dimensional data $\{a_2, a_3, a_8, a_9, a_{11}, a_{12}, d\}$ is presented in Figure 6.1. The multi-dimensional scaling plot of six dimensional data $\{a_2, a_3, a_8, a_9, a_{11}, a_{12}\}$ is presented in Figure 6.2.

Tables 6.18 and 6.19 contains the best quality indices of partitioning algorithms calculated for the set of selected 6 attributes and determined by them number of clusters. When there is the microalbuminuria included in the training data excepting CDbw tending to indicate 2 granules partitioning, the remaining indices suggest presence great number of granules in the data. The most indicated number of granules is 7, 9 and 10.

Table 6.18. Evaluation of the Best Clusterings of the Diabetes Data with Seven Attributes Selected Including the Decision Attribute Created by Algorithms: k-means, hcl and EM

Method	index	optimal value	nc
k-means	Dunn	0.63	5, 7
	DB	0.33	10
	SI	0.52	7
	CDbw	11.86	2
hcl	Dunn	0.62	5-7
	DB	0.29	10
	SI	0.53	10
	CDbw	11.86	2
EM	Dunn	0.53	3
	DB	0.04	9
	SI	0.37	6
	CDbw	13.04	3

Tables 6.20 and 6.21 hold evaluation of clusterings using SOSIG and DB-SCAN methods. In case of data with the decision attribute there are threshold values of the parameters s (SOSIG) and ε (DBSCAN), below which not enough number of input objects are clustered and over which all objects are joined in one cluster. The algorithms created 9 granules partitioning for narrow range of the parameters ($s \in [1.8, 2.1]$, $\varepsilon \in [0.4, 1.0]$), however the indices are greater for SOSIG result.

When excluding the microalbuminuria attribute from the data the number of granules present is decreasing. The most often distinguished clustering is the result composed of 4 granules. In Table 6.19 for CDbw index are placed 2 values of the index, since the second value, although smaller, were also distinguished by the index. The algorithms SOSIG and DBSCAN performed partitioning composed

Table 6.19. Evaluation of the Best Clusterings of the Diabetes Data with Six Attributes Selected Excluding the Decision Attribute Created by Algorithms: k-means, hcl and EM

Method	index	optimal value	nc
k-means	Dunn	0.77	4
	DB	0.25	8
	SI	0.55	4
	CDbw	17.11, 11.86	2 4
hcl	Dunn	0.63	4
	DB	0.16	9
	SI	0.57	5
	CDbw	17.11, 14.12	2 3
EM	Dunn	0.56	2,4
	DB	0.22	10
	SI	0.48	4
	CDbw	19.54, 10.93	2 4

Table 6.20. Evaluation of Clusterings of the Diabetes Data with Seven Attributes Selected Including the Decision Attribute Created by SOSIG and DBSCAN Algorithms

Method	index	$s < 1.8$	$s = 1.8$	$s = 2.1$	$s > 2.1$
SOSIG	Dunn	< 90%	0.86	0.86	all
	DB	objects	0.16	0.19	objects
	SI	clustered	0.64	0.63	in one
	CDbw		8.67	4.97	group
	index	$\varepsilon < 0.4$	$\varepsilon = 0.4$	$\varepsilon = 1.0$	$\varepsilon > 1.0$
DBSCAN	Dunn	< 90%	0.20	0.49	all
	DB	objects	0.50	0.30	objects
	SI	clustered	0.46	0.57	in one
	CDbw		1.66	1.90	group

of 6 granules. Assessment of the results are in Table 6.21. There is also a range of the parameters s and ε to cluster the data. The SOSIG algorithm granules are characterized of greater values of the assessment indices.

The best result is obtained by SOSIG algorithm (the best values of quality indices) and the result is taken for further analysis. The final granulation of diabetes data by SOSIG method is presented in Tables 6.22. Upper table concerns 7 dimensional data (the decision attribute included). The set is split in 9 granules of size 3 to 32. The first granule is greatest (32 objects). It is characterized by low to average level of HbA1c, not occurring any additional complications like Hypertension, Hypercholesterolemia or Hypertriglyceridemia. There is also absent microalbuminuria in this group. The following great granule contain 21 objects.

Table 6.21. Evaluation of Clusterings of the Diabetes Data with Six Attributes Selected Excluding the Decision Attribute Created by SOSIG and DBSCAN Algorithms

Method	index	$s < 2.2$	$s = 2.2$	$s = 2.8$	$s > 2.8$
SOSIG	Dunn	$< 90\%$	0.86	0.78	all
	DB	objects	0.17	0.19	objects
	SI	clustered	0.61	0.60	in one
	CDbw		12.68	12.87	group
	index	$\varepsilon < 0.3$	$\varepsilon = 0.3$	$\varepsilon = 1.0$	$\varepsilon > 1.0$
DBSCAN	Dunn	$< 90\%$	0.15	0.56	all
	DB	objects	0.32	0.25	objects
	SI	clustered	0.50	0.57	in one
	CDbw		4.15	14.83	group

Table 6.22. Composition of Granules of the Diabetes Data with 6 Attributes Selected Including (Upper Table) and Excluding (Bottom Table) the Decision Attribute Created by SOSIG Algorithm

Gr. lab.	No of obj.	a_2	a_3	a_8	a_9	a_{11}	a_{12}	d
1	32	[2,16]	(3,12]	[7.74-10.85]	0	0	0	0
2	9	(11,16]	[2,7)	[8.47-11.51]	1	0	0	1
3	5	(10,15)	(4,7)	[7.5-9.99]	1	0	0	0
4	6	(3,9]	(4,5)	[8.72-10.60]	0	1	0	0
5	3	(2,10)	(5,8)	[9.51-11.27]	0	1	1	0
6	13	[5,14)	(4,12]	[7.14-11.94]	0	1	0	1
7	21	(2,16]	(3,7)	[6.65-10.48]	0	0	0	1
8	7	(7,13]	(4,7)	[9.93-11.57]	0	1	1	1
9	3	(11,12]	(6,8)	[9.93-11.45]	1	1	1	1

Gr. lab.	No of obj.	a_2	a_3	a_8	a_9	a_{11}	a_{12}
1	53	[1,17]	(3,12)	[6.65-10.78]	0	0	0
2	14	(7,16]	(4,7)	[7.5-11.51]	1	0	0
3	10	(2,13]	(4,8)	[9.51-11.57]	0	1	1
4	20	(3,16]	(4,12]	[7.14-11.94]	0	1	0
5	3	(11,12]	(6,9]	[9.87-11.45]	1	1	1

Distinguished attributes are low to average of the disease duration (a_3) and low to average of HbA1c level. Similar to the previous group there are not present additional complication, however microalbuminuria appears. The third of the largest granules is composed of 13 objects representing patients of medium to late age of diabetes diagnosis, suffering from Hypercholesterolemia with microalbuminuria present. The remaining granules have considerably smaller size but are well distinguishable from the others owing to the features (see Table 6.22).

There are 6 granules of data distinguished by SOSIG in 6 dimensional diabetes data. One of them is considerable small (3 objects) comparing to the others (10-53 objects). The greatest granule consists objects regardless of value

Table 6.23. Semantically Described Granules Created by SOSIG with Applying Medical Norms (6 Attributes Selected)

Gr. lab.	No of obj.	a_2	a_3	a_8	a_9	a_{11}	a_{12}
1	53	all age	all time	below&norm	no	no	no
2	14	early school& adolescence	short&	whole range medium	yes	no	no
3	10	preschool& early school	short& medium	norm& over	no	yes	yes
4	20	all age	all time	whole range	no	yes	no
5	3	early school	medium	norm&over	yes	yes	yes

of attribute a_2 and a_3, with very small to average values of HbA1c and without Hypertension, Hypercholesterolemia and Hypertriglyceridemia. Following group contains 14 objects with average to very high value of a_2 (with medium to very late age of diabetes diagnosis), small to average disease duration a_3 and small to high HbA1c level. It occurs Hypertension in this group, however do not the remaining complications ($Ht = 0$, $Hg = 0$). The third granule contains 10 elements with low to high value of a_2 (with early to late age of diabetes diagnosis), small to average disease duration a_3 and average to high level of HbA1c. The granule is characterized by Hypertension not occurring, whereas problems of Hypercholesterolemia and Hypertriglyceridemia exist. The following granule of 20 objects is described by medium to very late age of diabetes diagnosis (a_2) regardless of the attributes a_3 and HbA1c. The patients from this group suffer from Hypercholesterolemia, whereas there is no problem with Hypertension and Hypertriglyceridemia. The smallest granule contains objects with medium age of illness diagnosis (a_2), average duration of diabetes and average to high HbAc1 level. The patients from the granule suffer from Hypertension and Hypercholesterolemia and Hypertriglyceridemia as well.

6.7 Conclusions

The approach presented in this chapter can be treated as an example of data mining for biological problems (one of challenging problems in data mining research [218]).

Part III

Complex Data and Complex Concepts

7 Mining Knowledge from Complex Data

7.1 Introduction

One important type of complex knowledge can occur when mining data from multiple relations. In most domains, the objects of interest are not independent of each other, and are not of a single type. For example in World Wide Web

- Text has a list structure. We consider sequences of words.
- HTML has a tree structure (nested tags).
- Hyperlinks have a graph structure (linked pages).

In fact, most real domains have combinations of different types of internal and external structure nested at multiple levels of abstraction. We need data mining systems that can soundly mine the rich structure of relations among objects, such as interlinked Web pages, social networks, metabolic networks in the cell, etc. Yet another important problem is how to mine non-relational data. For example described by formulas of first-order logic.

Approximation spaces are fundamental structures for the rough set approach [106, 108, 109, 110, 145]. In this chapter we show how the rough set approach can be used for mining knowledge from complex data.

In learning approximations of concepts, there is a need to choose a description language. This choice may limit the domains to which a given algorithm can be applied. There are at least two basic types of objects: structured and unstructured. An unstructured object is usually described by attribute-value pairs. For objects having an internal structure first order logic language is often used. Attribute-value languages have the expressive power of propositional logic. These languages sometimes do not allow for proper representation of complex structured objects and relations among objects or their components. The background knowledge that can be used in the discovery process is of a restricted form and other relations from the database cannot be used in the discovery process. Using first-order logic (or FOL for short) has some advantages over propositional logic [16, 30, 91]. First order logic provides a uniform and very expressive means of representation. The background knowledge and the examples, as well as the induced patterns, can all be represented as formulas in a first order language.

J. Stepaniuk: Rough - Gran. Comput. in Knowl. Dis. & Data Min., SCI 152, pp. 99–110, 2008.
springerlink.com © Springer-Verlag Berlin Heidelberg 2008

Unlike propositional learning systems, the first order approaches do not require that the relevant data be composed into a single relation but, rather, they can take into account data organized in several database relations with various connections existing among them.

The chapter is organized as follows. In Section 7.2 we discuss notions of relational learning. In Sections 7.3, 7.4 and 7.5 we consider application of rough set methods to discovery of interesting patterns expressed in a first order language. In Section 7.4 rough set methodology is used in the process of selecting relevant facts from background knowledge. The selection is based on constants occurring in positive and negative examples of a target relation. In Section 7.5 rough set methodology is used in the process of selecting literals which may be a part of a rule. In Section 7.6 we shortly discuss similarity measures combined with rough set methodology in classification and description of complex structured objects.

7.2 Relational Data Mining

Before moving on to the algorithm for learning of a set of rules, let us introduce some basic terminology from relational learning.

Relational learning algorithms learn classification rules for a concept [30, 91]. The program typically receives a large collection of positive and negative examples from real-world databases as well as background knowledge in the form of relations. Let p be a target predicate of arity m and $r_1, \ldots, r_l$ be background predicates, where $m, l > 0$ are given natural numbers. We denote the constants by $con_1, \ldots, con_n$, where $n > 0$. A term is either a variable or a constant. An atomic formula is of the form $p(t_1, \ldots, t_m)$ or $r_i(t_1, \ldots, t_m)$ where the t's are terms and $i = 1, \ldots, l$. A literal is an atomic formula or its negation. If a literal contains a negation symbol ($\neg$), we call it a negative literal, otherwise it is a positive literal. A clause is any disjunction of literals, where all variables are assumed to be universally quantified. The learning task for relational learning systems is as follows:

Input
a set X_{target}^+ of positive and a set X_{target}^- of negative training examples (expressed by literals without variables) for the target relation,
background knowledge (or BK for short) expressed by literals without variables and not including the target predicate.

Output
a set of $\xi \leftarrow \lambda$ (equivalently expressed in the form **if** λ **then** ξ) rules, where ξ is an atomic formula of the form $p(var_1^p, \ldots, var_m^p)$ with the target predicate p and λ is a conjunction of literals over background predicates $r_1, \ldots, r_l$, such that the set of rules satisfies the positive examples relatively to background knowledge.

We will adopt the lower and the upper approximations for subsets of the set of target examples. First, we define the coverage of a rule.

Definition 7.1. *A substitution is a mapping of variables to terms. The coverage of Rule, written Coverage(Rule), is the set of examples such that there exists* *a substitution giving values to all variables appearing in the rule (in such a way that all the occurrences of a given variable are replaced by the same term) and all literals of the rule are satisfied for this substitution.*

The set of the positive (negative) examples covered by *Rule* is denoted by $Coverage^+(Rule)$, $Coverage^-(Rule)$, respectively.

Remark 7.2. For any literal L, we obtain

$$Coverage(h \leftarrow b) = Coverage(h \leftarrow b \wedge L) \cup Coverage(h \leftarrow b \wedge \neg L).$$

Let $U = X^+_{target} \cup X^-_{target}$ and $Rule\_Set = \{Rule_1, \ldots, Rule_n\}$.

Definition 7.3. *For the set of rules $Rule\_Set$ and any example $x \in U$ the uncertainty function is defined by*

$$I_{Rule\_Set}(x) = \{x\} \cup \bigcup_{i=1}^{n} \{Coverage(Rule_i) : x \in Coverage(Rule_i)\}.$$

The lower and upper approximations may be defined as earlier but in this case they are equal to the forms presented in Remark 7.4.

Remark 7.4. For an approximation space $AS_{Rule\_Set} = (U, I_{Rule\_Set}, \nu_{SRI})$ and any subset $X \subseteq U$, the lower and the upper approximations are defined by

$$LOW(AS_{Rule\_Set}, X) = \{x \in U : I_{Rule\_Set}(x) \subseteq X\},$$

$$UPP(AS_{Rule\_Set}, X) = \{x \in U : I_{Rule\_Set}(x) \cap X \neq \emptyset\},$$

respectively.

7.3   From Complex Data into Attribute–Value Data

In this section we discuss the approach based on two steps. First, the data is transformed from first-order logic into decision table format by the iterative checking whether a new attribute adds any relevant information to the decision table. Next, the reducts and rules from reducts [106, 145, 189] are computed from the decision table obtained.

Data represented as a set of formulas can be transformed into attribute–value form. The idea of translation was inspired by LINUS and DINUS systems (see, e.g., [30]). We start with a decision table directly derived from the positive and negative examples of the target relation. Assuming that we have m-ary target predicate, the set U of objects in the decision table is a subset of $\{con_1, \ldots, con_n\}^m$. Decision attribute $d_p : U \rightarrow \{+, -\}$ is defined by the target predicate with possible values " $+$ " or " $-$ ". All positive and negative examples of the target predicate are now put into the decision table. Each example forms a separate row in the table. Then background knowledge is applied to the decision table. We determine all the possible applications of the background predicates to the arguments of the target relation. Each such application introduces a new Boolean attribute.

To analyze the complexity of the obtained data table, let us consider the number of condition attributes. Let A_{r_i} be a set of attributes constructed for every predicate symbol r_i, where $i = 1, \ldots, l$. The number of condition attributes in constructed data table is equal to $\sum_{i=1}^{l} card\,(A_{r_i})$ resulting from the possible applications of the l background predicates on the variables of the target relation. The cardinality of A_{r_i} depends on the number of arguments of target predicate p (denoted by m) and the arity of r_i. Namely, $card\,(A_{r_i})$ is equal to $m^{ar(r_i)}$, where $ar\,(r_i)$ is the arity of the predicate r_i. The number of condition attributes in obtained data table is polynomial in the arity m of the target predicate p and the number l of background knowledge predicates, but its size is usually so large that its processing is unfeasible. Therefore, one can check interactively if a new attribute is relevant, i.e., if it adds any information to the decision table and, next we add to the decision table only relevant attributes.

Two conditions for testing if a new attribute a is relevant are proposed:

1. $\gamma\left(AS_{B \cup \{a\}}, \{X_+, X_-\}\right) > \gamma\left(AS_B, \{X_+, X_-\}\right),$
   where X_+ and X_- denote the decision classes corresponding to the target concept. An attribute a is added to the decision table if this results in a growth of the positive region with respect to the attributes selected previously.
2. $Q_{DIS}(a) = \nu_{SRI}\left(X_+ \times X_-, \{(x,y) \in X_+ \times X_- : a\,(x) \neq a\,(y)\}\right) \geq \theta,$
   where $\theta \in [0,1]$ is a given real number. An attribute a is added to the decision table if it introduces some discernibility between objects belonging to different non-empty classes X_+ and X_-.

Each of these conditions can be applied to a single attribute before it is introduced to the decision table. If this attribute does not meet a condition, it should not be included into the decision table. The received data table is then analyzed by a rough set based systems. First, reducts are computed. Next, decision rules are generated.

Example 7.5. The problem with three binary predicates r_1, r_3, p and one unary predicate r_2 can be used to demonstrate the transformation of relational learning problem into attribute–value form. Suppose that there are the following positive and negative examples of a target predicate p :

$X_{target}^+ = \{p(1,2), p(4,1), p(4,2)\}, \quad X_{target}^- = \{\neg p(6,2), \neg p(3,5), \neg p(1,4)\}.$
Consider the background knowledge about relations, $r_1, r_2,$ and r_3 :
$r_1(5,1), r_1(1,2), r_1(1,4), r_1(4,1), r_1(3,1), r_1(2,6), r_1(3,5), r_1(4,2),$
$r_2(1), r_2(2), r_2(3), r_2(4), r_2(6), r_3(2,1), r_3(1,4), r_3(2,4),$
$r_3(2,5), r_3(3,2), r_3(3,5), r_3(5,1), r_3(5,3), r_3(2,6), r_3(4,2).$

We then transform the data into attribute–value form (decision table). In Table 7.1, a quality index Q_{DIS} of potential attributes is presented.

Using conditions introduced in this section some attributes will not be included in the resulting decision table. For example, the second condition with

Table 7.1. Quality Q_{DIS} of Potential Attributes

Symbol	Attribute	$Q_{DIS}(\bullet)$
a_1	$r_2(var_1)$	0
a_2	$r_2(var_2)$	0.33
a_3	$r_1(var_1, var_1)$	0
a_4	$r_1(var_1, var_2)$	0.33
a_5	$r_1(var_2, var_1)$	0.56
a_6	$r_1(var_2, var_2)$	0
a_7	$r_3(var_1, var_1)$	0
a_8	$r_3(var_1, var_2)$	0.56
a_9	$r_3(var_2, var_1)$	0.33
a_{10}	$r_3(var_2, var_2)$	0

Table 7.2. Resulting Decision Table $DT_{0.3}$ (t denotes true and f denotes false), Uncertainty Function and Rough Inclusion

(var_1, var_2)	a_2	a_4	a_5	a_8	a_9	d_p	$I_{A_{0.3}}(\bullet)$	$\nu_{SRI}(\bullet, X_+)$	$\nu_{SRI}(\bullet, X_-)$
$(1,2)$	t	t	f	f	t	+	$\{(1,2)\}$	1	0
$(4,1)$	t	t	t	f	t	+	$\{(4,1)\}$	1	0
$(4,2)$	t	t	f	t	t	+	$\{(4,2)\}$	1	0
$(6,2)$	t	f	t	f	t	-	$\{(6,2)\}$	0	1
$(3,5)$	f	t	f	t	t	-	$\{(3,5)\}$	0	1
$(1,4)$	t	t	t	t	f	-	$\{(1,4)\}$	0	1

$Q_{DIS}(\bullet) \geq \theta = 0.3$ would permit the following attribute set into the decision table: $A_{0.3} = \{a_2, a_4, a_5, a_8, a_9\}$.

Therefore, $DT_{0.3} = (U, A_{0.3} \cup \{d\})$ finally. We obtain two decision classes: $X_+ = \{(1,2), (4,1), (4,2)\}$ and $X_- = \{(6,2), (3,5), (1,4)\}$. For the obtained decision table we construct an approximation space $AS_{A_{0.3}} = (U, I_{A_{0.3}}, \nu_{SRI})$ such that the uncertainty function and the rough inclusion are defined in Table 7.2. Then, we can compute reducts and decision rules.

7.4 Selection of Relevant Facts

An approach presented in this section consists of the following steps:

1. Selection of potentially relevant facts from background knowledge.
2. Application of a relational learning system such as RSRL (see Section 7.5) to the selected formulas.

The selection is based on constants occurring in positive and negative examples of a target relation. The set of all constants occurring in a fact x is denoted by $CON(x)$. CON can be treated as a set valued attribute. A set of constants for a set of facts X is defined by $CON(X) = \bigcup_{x \in X} CON(x)$.

Training set reduction begins with determining the set of constants in all positive and negative examples for the target predicate. Such set is denoted as $CON(X_{target})$. We consider a data table $(U, \{CON\} \cup \{d\})$, where U is the set of all facts from background knowledge, $CON : U \rightarrow P(\{con_1, \ldots, con_n\})$, where $P(\{con_1, \ldots, con_n\})$ is the set of all subsets of constants and $d : U \rightarrow \{0, 1\}$. For every $x \in U$ we assume $d(x) = 1$ if and only if $CON(x) \subseteq CON(X_{target})$. The selections can be represented as lower and upper approximations of $X_{d=1} = \{x \in U : d(x) = 1\}$ in the family of approximation spaces $AS_{CON}^{f_{CON}} = \left(U, I_{CON}^{f_{CON}}, \nu_{SRI}\right)$, where

$$f_{CON}(CON(x), CON(x')) = w_1 * \frac{card(CON(x))}{card(CON(x) \cup CON(x'))} +$$

$$w_2 * \frac{card(CON(x'))}{card(CON(x) \cup CON(x'))} + \varepsilon$$

and w_1, w_2 and ε are parameters.

Definition 7.6. *Let* $AS_{CON}^{f_{CON}} = \left(U, I_{CON}^{f_{CON}}, \nu_{SRI}\right)$ *be an approximation space, where*

1. *U is the set of all formulas from background knowledge,*
2. *the uncertainty function* $I_{CON}^{f_{CON}}$ *is defined by*
   $x' \in I_{CON}^{f_{CON}}(x)$ *if and only if*

$$1 - \frac{card(CON(x) \cap CON(x'))}{card(CON(x) \cup CON(x'))} \leq f_{CON}(CON(x), CON(x')). \quad (7.1)$$

Any uncertainty function contributes to a different approximation space which results in different kinds of approximations that show different properties.

We then define two transformations $LOW : P(U) \rightarrow P(U)$ and $UPP : P(U) \rightarrow P(U)$ based on the lower and upper approximations in $AS_{CON}^{f_{CON}}$.

Starting with $X_{d=1}$ one can construct a sequence of approximations by constantly applying one of these transformations first on $X_{d=1}$ and then on the approximation resulting from the previous step.

Thus, the problem of selection is reduced to constantly applying upper (lower) approximation in the same approximation space to the upper (lower) approximation set obtained in the previous step.

The input data reduction problem is then defined as taking into account facts that are included in $LOW\left(AS_{CON}^{f_{CON}}, X_{d=1}\right)$. If this approximation appears to be too restrictive, which results in a bad quality of the discovered knowledge, then we consider $UPP\left(AS_{CON}^{f_{CON}}, X_{d=1}\right)$. If this does not meet our expectations, either, we proceed to consider the following approximations:

$$UPP\left(AS_{CON}^{f_{CON}}, UPP\left(AS_{CON}^{f_{CON}}, X_{d=1}\right)\right) \text{ and so on. We can stop when the}$$

approximation is sufficient to learn up in order to obtain a satisfactory definition of the target concept.

Since $X_{target} = X_{target}^+ \cup X_{target}^-$ (the union of positive and negative examples of the target relation) we may also consider separate approximations of sets

Table 7.3. Background Knowledge and Uncertainty Functions for $\varepsilon = 0.5$ and $\varepsilon = 0.25$

U	$x_\bullet$	CON	$I_{CON}^{0.5}(\bullet)$	$I_{CON}^{0.25}(\bullet)$
$k(5,1)$	x_1	$\{1,5\}$	$\{x_1, x_{11}, x_{24}\}$	$\{x_1, x_{24}\}$
$k(6,7)$	x_2	$\{6,7\}$	$\{x_2, x_{15}\}$	$\{x_2\}$
$k(1,2)$	x_3	$\{1,2\}$	$\{x_3, x_{11}, x_{12}, x_{17}\}$	$\{x_3, x_{17}\}$
$k(1,4)$	x_4	$\{1,4\}$	$\{x_4, x_5, x_{11}, x_{14}, x_{18}\}$	$\{x_4, x_5, x_{18}\}$
$k(4,1)$	x_5	$\{1,4\}$	$\{x_4, x_5, x_{11}, x_{14}, x_{18}\}$	$\{x_4, x_5, x_{18}\}$
$k(3,1)$	x_6	$\{1,3\}$	$\{x_6, x_{11}, x_{13}\}$	$\{x_6\}$
$k(2,6)$	x_7	$\{2,6\}$	$\{x_7, x_{12}, x_{15}, x_{27}\}$	$\{x_7, x_{27}\}$
$k(3,5)$	x_8	$\{3,5\}$	$\{x_8, x_{13}, x_{23}, x_{26}\}$	$\{x_8, x_{23}, x_{26}\}$
$k(3,7)$	x_9	$\{3,7\}$	$\{x_9, x_{13}, x_{25}\}$	$\{x_9, x_{25}\}$
$k(4,2)$	x_{10}	$\{2,4\}$	$\{x_{10}, x_{12}, x_{14}, x_{19}, x_{28}\}$	$\{x_{10}, x_{19}, x_{28}\}$
$h(1)$	x_{11}	$\{1\}$	$\{x_1, x_3, x_5, x_6, x_{11}, x_{17}, x_{18}, x_{24}\}$	$\{x_{11}\}$
$h(2)$	x_{12}	$\{2\}$	$\{x_3, x_7, x_{10}, x_{12}, x_{17}, x_{19}, x_{20}, x_{21}, x_{22}, x_{27}, x_{28}\}$	$\{x_{12}\}$
$h(3)$	x_{13}	$\{3\}$	$\{x_6, x_8, x_9, x_{13}, x_{22}, x_{23}, x_{25}, x_{26}\}$	$\{x_{13}\}$
$h(4)$	x_{14}	$\{4\}$	$\{x_4, x_5, x_{10}, x_{14}, x_{18}, x_{19}, x_{28}\}$	$\{x_{14}\}$
$h(6)$	x_{15}	$\{6\}$	$\{x_2, x_7, x_{15}, x_{27}\}$	$\{x_{15}\}$
$h(8)$	x_{16}	$\{8\}$	$\{x_{16}\}$	$\{x_{16}\}$
$r(2,1)$	x_{17}	$\{1,2\}$	$\{x_3, x_{11}, x_{12}, x_{17}\}$	$\{x_3, x_{17}\}$
$r(1,4)$	x_{18}	$\{1,4\}$	$\{x_4, x_5, x_{11}, x_{14}, x_{18}\}$	$\{x_4, x_5, x_{18}\}$
$r(2,4)$	x_{19}	$\{2,4\}$	$\{x_{10}, x_{12}, x_{14}, x_{19}, x_{28}\}$	$\{x_{10}, x_{19}, x_{28}\}$
$r(2,5)$	x_{20}	$\{2,5\}$	$\{x_{12}, x_{20}\}$	$\{x_{20}\}$
$r(2,7)$	x_{21}	$\{2,7\}$	$\{x_{12}, x_{21}\}$	$\{x_{21}\}$
$r(3,2)$	x_{22}	$\{2,3\}$	$\{x_{12}, x_{13}, x_{22}\}$	$\{x_{22}\}$
$r(3,5)$	x_{23}	$\{3,5\}$	$\{x_8, x_{13}, x_{23}, x_{27}\}$	$\{x_8, x_{23}, x_{26}\}$
$r(5,1)$	x_{24}	$\{1,5\}$	$\{x_1, x_{11}, x_{24}\}$	$\{x_1, x_{24}\}$
$r(7,3)$	x_{25}	$\{3,7\}$	$\{x_9, x_{13}, x_{25}\}$	$\{x_9, x_{25}\}$
$r(5,3)$	x_{26}	$\{3,5\}$	$\{x_8, x_{13}, x_{23}, x_{26}\}$	$\{x_8, x_{23}, x_{26}\}$
$r(2,6)$	x_{27}	$\{2,6\}$	$\{x_7, x_{12}, x_{15}, x_{27}\}$	$\{x_7, x_{27}\}$
$r(4,2)$	x_{28}	$\{2,4\}$	$\{x_{10}, x_{12}, x_{14}, x_{19}, x_{28}\}$	$\{x_{10}, x_{19}, x_{28}\}$

corresponding to X_{target}^+ and X_{target}^- which are added after the approximation process. This approach results in a more restrictive approximation.

We give an illustrative example of the proposed approach.

Example 7.7. Let $X_{target}^+ = \{z(1,2), z(4,1), z(4,2)\}$ be the set of positive examples and $X_{target}^- = \{\neg z(6,2), \neg z(3,5), \neg z(1,4)\}$ be the set of negative examples.

Let background knowledge be presented in the column labeled by U of the Table 7.3. All facts will be labeled by x with subscript (see the second column). The set of all constants occurring in a given fact from background knowledge is presented in column labeled by CON (see Table 7.3). We obtain $CON(X_{target}) = \{1,2,3,4,5,6\}$ and the set $X_{d=1}$ is equal to

$$\{x \in U : d(x) = 1\} = \{x \in U : CON(x) \subseteq CON(X_{target})\} =$$

$$= \{x_1, x_3, \ldots, x_8, x_{10}, \ldots, x_{15}, x_{17}, \ldots, x_{20}, x_{22}, x_{23}, x_{24}, x_{26}, x_{27}, x_{28}\}.$$

Table 7.4. Relevant Facts for Rule 7.2

X_{target}	$k(var_1, var_2)$	$h(var_2)$	$r(var_2, var_1)$	$z(var_1, var_2)$
$(1, 2)$	$+(x_3)$	$+(x_{12})$	$+(x_{17})$	$+$
$(4, 1)$	$+(x_5)$	$+(x_{11})$	$+(x_{18})$	$+$
$(4, 2)$	$+(x_{10})$	$+(x_{12})$	$+(x_{19})$	$+$
$(6, 2)$	$-$	$+(x_{12})$	$+(x_{27})$	$-$
$(3, 5)$	$+(x_8)$	$-$	$+(x_{26})$	$-$
$(1, 4)$	$+(x_4)$	$+(x_{14})$	$-$	$-$

For simplicity of presentation, we assume that in f_{CON} from Equation 7.1 the coefficient $w_1 = w_2 = 0$. Let us consider two examples $\varepsilon = 0.5$ and $\varepsilon = 0.25$. From the Equation 7.1 we obtain the uncertainty functions presented in columns labeled by $I_{CON}^{0.5}(\bullet)$ and $I_{CON}^{0.25}(\bullet)$.

We compute the lower and upper approximations for $\varepsilon \in \{0.25, 0.5, 0.75\}$.

$$LOW\left(AS_{CON}^{0.25}, X_{d=1}\right) = X_{d=1} = UPP\left(AS_{CON}^{0.25}, X_{d=1}\right) = \{x_1, x_3, x_4, x_5, x_6,$$

$$x_7, x_8, x_{10}, x_{11}, x_{12}, x_{13}, x_{14}, x_{15}, x_{17}, x_{18}, x_{19}, x_{20}, x_{22}, x_{23}, x_{24}, x_{26}, x_{27}, x_{28}\}.$$

$$LOW\left(AS_{CON}^{0.5}, X_{d=1}\right) = \{x_1, x_3, x_4, x_5, x_6,$$

$$x_7, x_8, x_{10}, x_{11}, x_{12}, x_{14}, x_{17}, x_{18}, x_{19}, x_{20}, x_{22}, x_{23}, x_{24}, x_{26}, x_{27}, x_{28}\}.$$

$$UPP\left(AS_{CON}^{0.5}, X_{d=1}\right) = U - \{x_{16}\}.$$

$$LOW\left(AS_{CON}^{0.75}, X_{d=1}\right) = \{x_1, x_4, x_5, x_{11}, x_{14}, x_{18}, x_{23}, x_{24}\}.$$

$$UPP\left(AS_{CON}^{0.75}, X_{d=1}\right) = U - \{x_{16}\}.$$

For $\varepsilon = 0.25$ we do not obtain any reduction of the set $X_{d=1}$.

For $\varepsilon = 0.75$ the lower approximation reduces the number of facts in background knowledge. The upper approximation is almost equal to U, thus, further iteration of upper approximation seems to be unnecessary in this case as well as for $\varepsilon = 0.5$.

Let us observe that for the illustrative rule

$$z(var_1, var_2) \leftarrow k(var_1, var_2) \wedge h(var_2) \wedge r(var_2, var_1) \qquad (7.2)$$

the facts from $X_{d=1} - LOW\left(AS_{CON}^{0.5}, X_{d=1}\right) = \{x_2, x_9, x_{13}, x_{15}, x_{16}, x_{21}, x_{25}\}$ seems to be irrelevant (see Table 7.4).

7.5   The Rough Set Relational Learning Algorithm

In this section we recall the RSRL (**R**ough **S**et **R**elational **L**earning) algorithm [191]. Some preliminary versions of this algorithm were presented in [194, 195].

Input to RSRL consists of relations defined extensively as sets of tuples of constants. RSRL also needs examples of tuples that do not belong to the target relation (negative examples). A type of approximation should also be specified. RSRL uses a covering approach similar to FOIL (see references in [30]).

To select the most promising literal from the candidates generated at each step, RSRL considers the performance of the rule over the training data. The evaluation function $card(R(L, NewRule))$ used by RSRL to estimate the utility of adding a new literal is based on the numbers of discernible positive and negative examples before and after adding the new literal (see Figure 7.1).

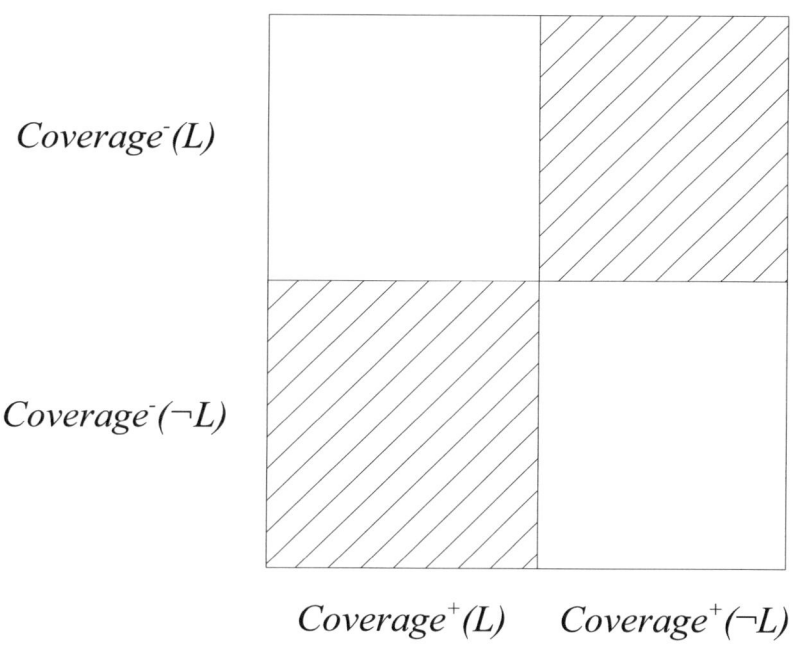

$Coverage^-(L)$

$Coverage^-(\neg L)$

$Coverage^+(L)$ $Coverage^+(\neg L)$

Fig. 7.1. The set of discernible positive and negative examples before and after adding the new literal L is equal to the union of two Cartesian products ($Coverage^+(L) \times Coverage^-(\neg L)) \cup (Coverage^+(\neg L) \times Coverage^-(L))$

Some modification of the algorithm RSRL were presented in [195]. The modified algorithm generates rules as the original RSRL but its complexity is lower because it performs operations on the cardinalities of sets without computing the sets.

7.6   Similarity Measures and Complex Objects

In [53, 55] an algorithm for classification of complex structured objects is proposed. The algorithm is designed for data expressed in a first-order logic language. Such a data set consists of sets of positive and negative examples of a

---

Algorithm 1. RSRL Algorithm

---

input   : $Target\_predicate, BK, X^+_{target} \cup X^-_{target}, app$ //where $Target\_predicate$
is a target predicate with a set X^+_{target} of positive examples and a set
X^-_{target} of negative examples, BK is a background knowledge, app is a
type of approximation ($app \in \{lower, upper\}$).

output: $Learned\_rules$ //where $Learned\_rules$ is a set of rules for "positive
decision class".

$Pos \longleftarrow X^+_{target}$;
$Learned\_rules \longleftarrow \emptyset$;
while $Pos \neq \emptyset$ **do**

> Learn a NewRule;
> $NewRule \longleftarrow$ most general rule possible;
> $R \longleftarrow Pos \times X^-_{target}$;
> **while** $R \neq \emptyset$ **do**
>
> > $Candidate\_literals \longleftarrow$ generated candidates; // RSRL generates
> > candidate specializations of $NewRule$ by considering a new literal L that
> > fits one of the following forms:
> >
> > - $r(var_1, \ldots, var_s)$, where at least one of the variable var_i in the created
> >   literal must already exist in the positive literals of the rule;
> > - the negation of the above form of literal;
> >
> > $Best\_literal \longleftarrow \arg\max_{L \in Candidate\_literals} card(R(L, NewRule))$; //
> > (the explanation of $R(L, Rule)$ is depicted in Figure 7.1)
> > **if** $R(Best\_literal, NewRule) = \emptyset$ or ($app = upper$ and ($NewRule \neq$ most
> > general rule possible)
> > and $Coverage^+(NewRule) \neq Coverage^+(NewRule \wedge Best\_literal)$) **then**
> >
> > > | exit while;
> >
> > **end**
> > Add $Best\_literal$ to $NewRule$ preconditions; //Add a new literal to
> > specialize $NewRule$;
> > **if** $Coverage^-(NewRule) = \emptyset$ **then**
> >
> > > | exit while;
> >
> > **end**
> > $R := R \setminus R(Best\_literal, NewRule)$;
>
> **end**
> $Learned\_rules \longleftarrow Learned\_rules \cup \{NewRule\}$;
> $Pos \longleftarrow Pos \setminus Coverage^+(NewRule)$;

end

---

target relation, and a background knowledge including literals obtained on the
base of other relations. Target examples are understood as complex structured
objects. The task of the algorithm is to find a pattern enabling to distinguish
positive examples from the negative ones. In presented approach, the pattern is
represented by a similarity degree of target examples. Some similarity measure
is applied to target examples in order to compare them. Target examples are as
similar as background literals related to the target examples. In order to find
a similarity degree of examples, a set of candidates being real numbers in the
range of $[0, 1]$ is considered. For each of them, the algorithm checks how many

positive and negative examples are similar to other positive examples. We select a degree for which the number of similar positive (negative) examples is highest (lowest). In the process of searching for a similarity degree, some notions of rough set methodology are applied. The author uses an approximation space in which the uncertainty function is constructed on the base of a similarity measure proposed in the paper. The lower and the upper approximations defined in the approximation space are applied to target examples. The lower or the upper approximation of a set of positive examples are considered. The type of approximation can be fixed by a user. An example belongs to the approximation if it is similar at least to the degree under consideration to positive examples. The approximation is computed for each degree of a set of candidates. A degree found over a training data is used in classification of new examples. An example is classified as positive if it belongs to the approximation computed with respect to the degree, and it is classified as negative otherwise. The paper [53] also includes results of some experiments performed by the algorithm. Two data sets were used in the experiments. The first one is related to the document understanding and describes components of single page documents. The second one is related to the family relations. In general, the algorithm obtained a high percentage of examples classified correctly.

An essential advantage of the approach presented in [53, 55] is the size of the pattern generated over a training data. Regardless of a data set, the pattern is one real number in the range of $[0, 1]$. However, an algorithm has some drawback, namely the training data set is needed in the process of classification of new examples. On the base of the results of experiments, one can observe that effectiveness of the approach depends on selection of the parameters and the type of approximation. A measure applied to compute a similarity degree may have a significant influence on the results of experiments.

In [54, 55] several similarity measures on complex structured objects are investigated. The measures are designed for data expressed in a first-order logic language. The goal of the proposed similarity measures is to distinguish positive examples from the negative ones. It is based on the assumption that positive examples are more similar to each other than to the negative ones. Some measures proposed in [54] are based on simple measures known in literature. Other more advanced measures are based on distance ones, and some of them use a notion of permutation. Similarity measures are the base of an algorithm for description and classification of objects. The algorithm proposed in [54] transforms data expressed in the first-order logic language into a decision table. The decision table returned by the algorithm is of the following form:

- objects represent target examples,
- values of condition attributes are computed by applying similarity measures (of examples) or functions using the similarity measures,
- the values of decision attribute corresponds to classes of positive and negative examples.

The approach is similar to approaches based on propositionalization [30]. However, an attribute is constructed in a different way than in case of a typical propositionalization method. Data stored in such a table can be analyzed by any appropriate attribute-value learner. In presented approach, rough set methods are used in order to induce decision rules from the table. The rules are treated as a description of target examples, and they are used in classification of new examples. In experiments, the author uses a decision table consisting of one condition attribute. The attribute expresses for each example of the table its similarity to positive examples. Two methods of construction of such an attribute are proposed. The paper also includes results of some experiments performed by the algorithm. Similarly as in [53], two data sets were used in the experiments. The first one is related to the document understanding and describes components of single page documents. The second one is related to the family relations. In general, the algorithm obtained a high percentage of examples classified correctly. One can conclude that effectiveness of the algorithm depends on factors such as:

- a way of computation of subsets of background literals related to target examples,
- a similarity measure applied to distinguishing target examples,
- a method applying similarity measures to generation of patterns describing positive (and alternatively negative) examples.

Some goals of future research are an extension of similarity measures to data with different types and a consideration of a decision table with more than one condition attribute.

7.7   Conclusions

Four cases of application of rough set methods to discovery of interesting patterns from relational data are described. The first case presented in this chapter is based on translation of the first order data into decision table format. The second case is based on selection of potentially relevant facts from background knowledge. The third case is based on the algorithm RSRL for the first order rules generation. We showed that approximation spaces are basic structures for knowledge discovery from complex data (multi-relational data). The fourth case is based on combining similarity measures with rough set methods.

Furthermore, the presented approach can be treated as a step towards the understanding of rough set methods in the problem of mining complex knowledge from complex data (see challenging problems in data mining research [218]).

8 Complex Concept Approximations

The rough set approach was further developed to deal with more compound granules than elementary granules. In this chapter, we present a methodology for modeling of compound granules using the rough set approach. The methodology is based on operations on information systems. There are two basic steps in such a modeling. In the first step, new granules are constructed from objects represented by granules in some already constructed information systems. These new granules are used as objects in the new constructed information systems. In the second step the features of the new granules are added. This approach can be used for modeling, e.g., compound granules in spatio-temporal reasoning.

This chapter is structured as follows. In Section 8.1 we recall the definition of information granulation. We also discuss systems of granules and examples of granules. In Section 8.2 we investigate granules in multiagent systems. In Section 8.3 we discuss modeling of compound granules based on information systems. In Section 8.4 we discuss rough-fuzzy granules.

8.1 Information Granulation and Granules

The concept of information granulation is rooted in several papers starting with [220] in which the concepts of a linguistic variable and granulation were introduced. Information granulation is performed on granules. In this chapter we assume that any granule is a pair:

$$(name, \ content), \tag{8.1}$$

where *name* is the granule name and *content* describes details of the granule construction together with their meaning (semantics).

In many examples, the granule names (labels) are formulas from some language and the granule contents are interpreted as the semantics of such formulas. In other examples, granule contents can have more compound structures defined by some other granules. For example, one can consider a granule representing a cluster

$$(patient\_cluster, \ cluster), \tag{8.2}$$

J. Stepaniuk: Rough - Gran. Comput. in Knowl. Dis. & Data Min., SCI 152, pp. 111–131, 2008.
springerlink.com

where *patient_cluster* is the name of a patient cluster in a medical database having similar symptoms and *cluster* consists of a cluster definition including the cluster construction and its semantics. Let us consider one more example:

$$(safe, \ classifier), \tag{8.3}$$

where *safe* is a vague concept describing that the situation of the road is *safe* and *classifier* is the induced approximation of the vague concept *safe*. Each classifier can be treated as a granule with *name* describing the classifier construction and *content* describing the classifier semantics. The granule presented in the last example describes, in a sense, the *meaning* of the vague concept *safe* relative, e.g., to an agent implemented in a computer system.

In the following sections, we present some systems of granules making it possible to describe construction of different kinds of basic granules.

8.1.1 Granule Systems

In this section, we present a basic notion for our approach, i.e., granule system. Any such system S consists of a set of granules G. Moreover, a family of relations with the intended meaning *to be a part to a degree* between granules is distinguished. The degree structure is described by a relation *to be an exact part*. More formally, a granule system is any tuple

$$S = (G, H, <, \{\nu_p\}_{p \in H}, size) \tag{8.4}$$

where

1. G is a non-empty set of granules;
2. H is a non-empty set of granule inclusion degrees with a binary relation $<$ (usually a strict partial order) which defines on H a structure used to compare the degrees;
3. $\nu_p \subseteq G \times G$ is a binary relation *to be a part to a degree at least p* between granules from G, called *rough inclusion*;
4. $size : G \longrightarrow R_+$ is the granule size function, where R_+ is the set of nonnegative reals.

In constructing of granule systems it is necessary to give a constructive definition of all their components.

In particular, one should specify how more compound granules are defined from already defined granules or given elementary granules. Usually, the set of granules is defined as the least set generated from distinguished elementary granules by some operations on the granules. In the following sections, we discuss several examples of such operations.

One can consider the following examples of formulas defining elementary granules:

1. a set of descriptors (selectors) of the form (a, v) where $a \in A$ and $v \in V_a$ for some finite attribute set A and value sets V_a;
2. a set of descriptor conjunctions.

In the standard rough set model granules are corresponding to indiscernibility classes of an equivalence relation.

Examples of complex granules are tolerance granules created by means of similarity (tolerance) relation between elementary granules, decision rules or sets of decision rules.

Notice that the existing measures of inclusion should be extended on more compound granules. Strategies for these extensions are selected so that constructed granules allow us to make a progress in constructing the target granules. In the following sections, we outline some of these issues in modeling of granules.

8.1.2   Name and Content: Syntax and Semantics

In this section, we present examples of granules with names defined by some formulas and with contents defined by semantics of these formulas.

Formulas are used to express properties of objects. Hence, we assume that together with a given information system there are defined

- a set of formulas Φ over some language,
- semantics Sem of formulas from Φ, i.e., a function from Φ into the power set $P(U)$.

Let us consider an example [106]. We define a language L_{IS} used for elementary granule description, where $IS = (U, A)$ is an information system. The syntax of L_{IS} is defined recursively by

1. $(a \; in \; V) \in L_{IS}$, for any $a \in A$ and $V \subseteq V_a$.
2. If $\alpha \in L_{IS}$ then $\neg \alpha \in L_{IS}$.
3. If $\alpha, \beta \in L_{IS}$ then $\alpha \wedge \beta \in L_{IS}$.
4. If $\alpha, \beta \in L_{IS}$ then $\alpha \vee \beta \in L_{IS}$.

The semantics of formulas from L_{IS} with respect to an information system IS is defined recursively by

1. $Sem_{IS}(a \; in \; V) = \{x \in U : a(x) \in V\}$.
2. $Sem_{IS}(\neg \alpha) = U - Sem_{IS}(\alpha)$.
3. $Sem_{IS}(\alpha \wedge \beta) = Sem_{IS}(\alpha) \cap Sem_{IS}(\beta)$.
4. $Sem_{IS}(\alpha \vee \beta) = Sem_{IS}(\alpha) \cup Sem_{IS}(\beta)$.

We now present the syntax and the semantics of examples of granules. These granules are constructed by taking collections of already specified granules. They are comprise parameters which can be adjusted in applications. In the following sections we discuss some other kinds of operations on granules as well as the inclusion and closeness relations for such granules.

Let us note that any granule g formally can be defined by the granules syntax $Syn(g)$ and semantics $Sem(g)$. However, for simplicity of notation we often use only one component of the granules to denote it.

8.1.3   Examples of Granules

Elementary granules. In an information system $IS = (U, A)$, elementary granules are defined by $EF_B(x)$, where EF_B is a conjunction of selectors (descriptors) of the form $a = a(x)$, $B \subseteq A$ and $x \in U$. For example, the meaning of an elementary granule $a = 1 \wedge b = 1$ is defined by
$Sem_{IS}(a = 1 \wedge b = 1) = \{x \in U : a(x) = 1 \ \& \ b(x) = 1\}$.

Thus, in the system

$$S_B = (G_B, H, <, \{\nu_p\}_{p \in H}, size) \tag{8.5}$$

of elementary granules G_B is a set of conjunctions of selectors, $H = [0, 1]$ and $\nu_p(EF_B, EF'_B)$ if and only if

$$\frac{card\,(Sem_{IS}\,(EF_B) \cap Sem_{IS}\,(EF'_B))}{card\,(Sem_{IS}\,(EF_B))} \geq p$$

The number of conjuncts in the granule can be taken as the granule size and it is one of parameters to be tuned, e.g., by the dropping condition technique used in machine learning.

One can extend the set of elementary granules assuming that if α is any Boolean combination of descriptors over A, then $(\overline{B}\alpha)$ and $(\underline{B}\alpha)$ define syntax of elementary granules too, for any $B \subseteq A$.

Sequences of granules. Let us assume that S is a sequence of granules and the semantics $Sem_{IS}(\bullet)$ in IS of its elements have been defined. We extend $Sem_{IS}(\bullet)$ on S by

$$Sem_{IS}(S) = \{Sem_{IS}(g)\}_{g \in S}.$$

Example 8.1. Granules defined by rules in information systems are examples of sequences of granules. Let IS be an information system and let (α, β) be a new granule received from the rule **if** α **then** β where α, β are elementary granules of IS. The semantics $Sem_{IS}((\alpha, \beta))$ of (α, β) is the pair of sets

$$(Sem_{IS}(\alpha), Sem_{IS}(\beta)).$$

If the right hand sides of rules represent decision classes, then the number of conjuncts on the left hand sides is one of the parameters to be adjusted during classifier construction. A typical goal is to search for minimal (or less than minimal) number of such conjuncts (corresponding to the largest generalization) which still guarantee the satisfactory degree of inclusion in a decision class.

Sets of granules. Let us assume that a set G of granules and the semantics $Sem_{IS}(\bullet)$ in IS for granules from G have been defined. We extend $Sem_{IS}(\bullet)$ on the family of sets $H \subseteq G$ by $Sem_{IS}(H) = \{Sem_{IS}(g) : g \in H\}$. One can consider as a parameter of any such granule its cardinality or its size (e.g., the length of such granule representation). In the first case, a typical problem is to search in a given family of granules for a granule of the smallest cardinality sufficiently close to a given one.

Example 8.2. One can consider granules defined by sets of rules. Assume that there is a set of rules $Rule\_Set = \{(\alpha_i, \beta_i) : i = 1, \ldots, k\}$. The semantics of $Rule\_Set$ is defined by

$$Sem_{IS}(Rule\_Set) = \{Sem_{IS}((\alpha_i, \beta_i)) : i = 1, \ldots, k\}.$$

The above mentioned searching problem for a set of granules corresponds in the case of rule sets to searching for the simplest representation of a given rule collection by another set of rules (or a single rule) sufficiently close to the collection.

Example 8.3. Let us consider a set G of elementary granules – describing possible situations together – with decision table DT_α representing decision tables for any situation $\alpha \in G$. Assume $Rule\_Set(DT_\alpha)$ to be a set of decision rules generated from decision table DT_α (e.g., in the minimal form). Now let us consider a new granule

$$\{(\alpha, Rule\_Set(DT_\alpha)) : \alpha \in G\}$$

with semantics defined by

$$\{Sem_{DT}((\alpha, Rule\_Set(DT_\alpha))) : \alpha \in G\} = \\ \{(Sem_{IS}(\alpha), Sem_{DT}(Rule\_Set(DT_\alpha))) : \alpha \in G\}.$$

An example of a parameter to be tuned is the number of situations represented in such granule. A typical task is to search for a granule with the minimal number of situations creating together with the sets of rules, corresponding to them, a granule sufficiently close to the original one.

Extension of granules defined by tolerance relation
Now we present examples of granules obtained by application of a tolerance relation (i.e., reflexive and symmetric relation; for more information see, e.g., [145]).

Example 8.4. One can consider extension of elementary granules defined by tolerance relation. Let $IS = (U, A)$ be an information system and let τ be a tolerance relation on elementary granules of IS. Any pair $(\tau : \alpha)$ is called a *τ-elementary granule*. The semantics $Sem_{IS}((\tau : \alpha))$ of $(\tau : \alpha)$ is the family $\{Sem_{IS}(\beta) : (\beta, \alpha) \in \tau\}$. Parameters to be tuned in searching for relevant tolerance granule can be its support (represented by the number of supporting it objects) and its degree of its inclusion (or closeness) in some other granules as well as parameters specifying the tolerance relation.

Example 8.5. Let us consider granules defined by rules of a tolerance information systems [145]. Let $IS = (U, A)$ be an information system and let τ be a tolerance relation on elementary granules of IS. If **if** α **then** β is a rule in IS then the semantics of a new granule $(\tau : \alpha, \beta)$ is defined by $Sem_{IS}((\tau : \alpha, \beta)) = Sem_{IS}((\alpha, \tau)) \times Sem_{IS}((\beta, \tau))$. Parameters to be tuned are the same as in the case of granules being sets of more elementary granules as well as parameters of the tolerance relation.

Example 8.6. We consider granules defined by sets of decision rules correspond-ing to a given evidence in tolerance decision tables. Let $DT = (U, A, d)$ be a decision table and let τ be a tolerance on elementary granules of $IS = (U, A)$. Now, any granule $(\alpha, Rule\_Set\,(DT_\alpha))$ can be considered as a representative for the granule cluster

$$(\tau : (\alpha, Rule\_Set\,(DT_\alpha)))$$

with the semantics

$$Sem_{DT}\,((\tau : (\alpha, Rule\_Set\,(DT_\alpha)))) = \\ \{Sem_{DT}\,((\beta, Rule\_Set\,(DT_\beta))) : (\beta, \alpha) \in \tau\}\,.$$

One can see that the considered case is a special case of the granules from Example 8.3 with G defined by a tolerance relation.

8.2   Granules in Multiagent Systems

Granules are involved in many tasks of approximate reasoning in multiagent systems [78]. Among them are

1. understanding granules used by other agents;
2. interaction of granules in searching for patterns used for compound concepts approximation;
3. discovery of new granules in interaction with environments used for predic-tion of behavior.

We will present several compound granules used in solving such tasks.

We begin from a short discussion on approximation spaces, i.e., granules used in concept approximation. Approximation spaces can be used by a given agent for approximating concepts used by another agent [135, 152]. In the simplest case, as the result of such an interaction, a decision table is obtained. This table is then used for concept approximation. We show that concept approximation on an extension of a given sample of objects can be treated as a searching task for an extension of approximation space. We also outline applications of granules in compound concept approximation where compound hierarchical patterns (hier-archical pattern granules) for approximation of such concepts are constructed by interaction of simpler patterns (pattern granules). Granulation of such patterns is used in searching for concept approximations (data models) with (sub-)minimal size and with satisfactory quality. It is worthwhile to mention that granulation can also be applied to rules used for patterns construction. In approximation of compound concepts we propose to use a domain ontology that makes the searching for compound concept approximation feasible [8, 9, 10, 154, 200]. In the ontology [169] (vague) concepts and local dependencies between them are specified. Global dependencies can be derived from local dependencies. Such derivations can be used as hints in searching for relevant compound patterns (granules) in approximation of more compound concepts from the ontology. The ontology approximation problem is one of the fundamental problems related to approximate reasoning in distributed environments. One should construct (in a given language that is different from the ontology specification language) not only approximations of concepts from ontology but also vague dependencies specified in the ontology. It is worthwhile to mention that an ontology approximation should be induced on the basis of incomplete information about concepts and dependencies specified in the ontology. Granule calculi based on rough sets have been proposed as tools making it possible to solve this problem. Vague dependencies have vague concepts in premises and conclusions. The approach to approximation of vague dependencies based only on degrees of closeness of concepts from dependencies and their approximations (classifiers) is not satisfactory for approximate reasoning. Hence, more advanced approach should be developed. Approximation of any vague dependency is a method which allows for any object to compute the arguments "for" and "against" its membership to the dependency's conclusion on the basis of the analogous arguments relative to the dependency's premises. Any argument is a compound granule (compound pattern). Arguments are fused by local schemes (production rules) discovered from data. Further fusions are possible through composition of local schemes, called approximate reasoning schemes (AR schemes) (see, e.g., [10, 104, 212]). To estimate the degree to which (at least) an object belongs to concepts from ontology the arguments "for" and "against" those concepts are collected and next a conflict resolution strategy is applied to them to predict the degree. This inference process is analogous to the inference process used in fuzzy logic [65] with numerical degree of membership functions. In the considered case, the numerical values are substituted by arguments "for" and "against" and the fuzzification is replaced by rules defining how the arguments from the left hand sides of dependencies are transformed to arguments for the concepts on the right hand sides. The defuzzification is substituted by conflict resolution strategy. In the discussed approach it is assumed that the rules are discovered from data and domain knowledge.

The performed experiments based on approximation of concept ontology (see, e.g., [3, 4, 104, 145, 152, 154, 158, 193]) showed that domain knowledge enables to discover relevant patterns in sample objects for compound concept approximation. Our approach to compound concept approximation and approximate reasoning about compound concepts is based on the rough-granular approach.

For modeling computations of multiagent systems more compound granules are needed. Let us observe that granule systems themselves can also be treated as granules. For example, each agent can be represented by a granule system. Moreover, in granular computations modeling the behavior of multiagent systems some specific operations on granules representing granule systems should be defined. There are several reasons for introducing such operations. For example,

1. granule systems of agents should be adaptively changed in interaction of agents with environments;
2. during coalition formation by a team of agents their granule systems should be fused into a new granule system relevant for the new coalition.

Other compound granules are needed for reasoning about behavior of agents in multiagent systems.

8.2.1 Rough Set Approach to Concept Approximation

In this section we consider the problem of approximation of concepts over a universe U^∞ (concepts that are subsets of U^∞). We assume that the concepts are perceived only through some subsets of U^∞, called samples. This is a typical situation in the machine learning, pattern recognition, or data mining approaches [47, 69]. We explain the rough set approach to induction of concept approximations using the generalized approximation spaces of the form $AS = (U, I_\#, \nu_\$)$ and an extension operation of such approximation spaces.

Let $U \subseteq U^\infty$ be a finite sample. By $\Pi_U : P(U^\infty) \to P(U)$ we denote a perception function from $P(U^\infty)$ into $P(U)$ defined by $\Pi_U(C) = C \cap U$ for any concept $C \subseteq U^\infty$ (see Figure 8.1).

Let $AS = (U, I, \nu)$ be an approximation space over the sample U. The problem we consider is how to extend the approximations of $\Pi_U(C)$ defined by AS to approximation of C over U^∞. We show that the problem can be described as

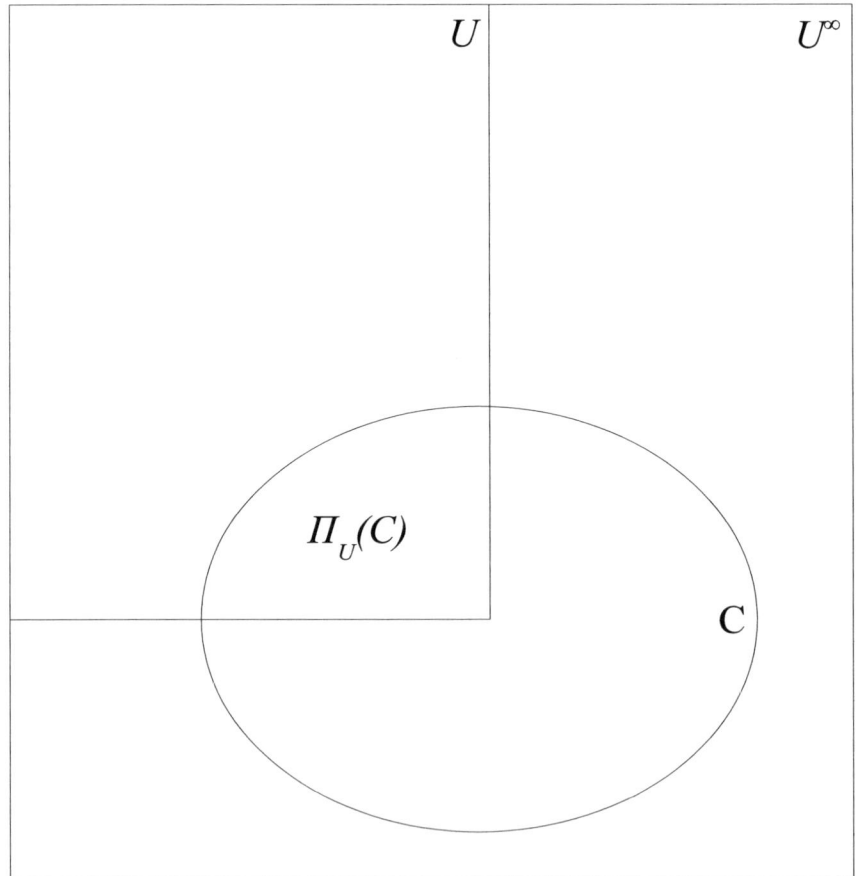

Fig. 8.1. Perception Function

searching for an extension $AS_C = (U^\infty, I_C, \nu_C)$ of the approximation space AS, relevant for approximation of C. This requires to show how to extend the inclusion function ν from subsets of U to subsets of U^∞ that are relevant for the approximation of C. Observe that for the approximation of C it is enough to induce the necessary values of the inclusion function ν_C without knowing the exact value of $I_C(x) \subseteq U^\infty$ for $x \in U^\infty$.

Let AS be a given approximation space for $\Pi_U(C)$ and let us consider a language L in which the neighborhood $I(x) \subseteq U$ is expressible by a formula $pat(x)$, for any $x \in U$. It means that $I(x) = \|pat(x)\|_U \subseteq U$, where $\|pat(x)\|_U$ denotes the meaning of $pat(x)$ restricted to the sample U. In case of rule based classifiers patterns of the form $pat(x)$ are defined by attribute value vectors.

We assume that for any new object $x \in U^\infty \setminus U$ we can obtain (e.g., as a result of sensor measurement) a pattern $pat(x) \in L$ with semantics $\|pat(x)\|_{U^\infty} \subseteq U^\infty$. However, the relationships between granules over U^∞ like sets: $\|pat(x)\|_{U^\infty}$ and $\|pat(y)\|_{U^\infty}$, for different $x, y \in U^\infty$, are, in general, known only if they can be expressed by relationships between the restrictions of these sets to the sample U, i.e., between sets $\Pi_U(\|pat(x)\|_{U^\infty})$ and $\Pi_U(\|pat(y)\|_{U^\infty})$.

The set of patterns $\{pat(x) : x \in U\}$ is usually not relevant for approximation of the concept $C \subseteq U^\infty$. Such patterns are too specific or not enough general, and can directly be applied only to a very limited number of new objects. However, by using some generalization strategies, one can search, in a family of patterns definable from $\{pat(x) : x \in U\}$ in L, for such new patterns that are relevant for approximation of concepts over U^∞. Let us consider a subset $PATTERNS(AS, L, C) \subseteq L$ chosen as a set of pattern candidates for relevant approximation of a given concept C. For example, in case of rule based classifier one can search for such candidate patterns among sets definable by subsequences of attribute value vectors corresponding to objects from the sample U. The set $PATTERNS(AS, L, C)$ can be selected by using some quality measures checked on meanings (semantics) of its elements restricted to the sample U (like the number of examples from the concept $\Pi_U(C)$ and its complement that support a given pattern). Then, on the basis of properties of sets definable by these patterns over U, we induce approximate values of the inclusion function ν_C on subsets of U^∞ definable by any of such pattern and the concept C.

Next, we induce the value of ν_C on pairs (X, Y) where $X \subseteq U^\infty$ is definable by a pattern from $\{pat(x) : x \in U^\infty\}$ and $Y \subseteq U^\infty$ is definable by a pattern from $PATTERNS(AS, L, C)$.

Finally, for any object $x \in U^\infty \setminus U$ we induce the approximation of the degree $\nu_C(\|pat(x)\|_{U^\infty}, C)$ applying a conflict resolution strategy $Conflict\_res$ (a voting strategy, in case of rule based classifiers) to two families of degrees:

$$\{\nu_C(\|pat(x)\|_{U^\infty}, \|pat\|_{U^\infty}) : pat \in PATTERNS(AS, L, C)\}, \qquad (8.6)$$

$$\{\nu_C(\|pat\|_{U^\infty}, C) : pat \in PATTERNS(AS, L, C)\}. \qquad (8.7)$$

Values of the inclusion function for the remaining subsets of U^∞ can be chosen in any way – they do not have any impact on the approximations of C. Moreover, observe that for the approximation of C we do not need to know the exact values of the uncertainty function I_C – it is enough to induce the values of the inclusion function ν_C. Observe that the defined extension ν_C of ν to some subsets of U^∞ makes it possible to define an approximation of the concept C in a new approximation space AS_C.

Observe that one can also follow principles of Bayesian reasoning and use degrees of ν_C to approximate C.

In this way, the rough set approach to induction of concept approximations can be explained as a process of inducing a relevant approximation space.

Any approximation space can be treated as a compound granule labeled by many parameters such as attribute sets defining the neighborhoods, rough inclusions, neighborhood size measures, parameters of patterns used for estimation of the extension of the rough inclusion. One can define the quality of a given approximation space relative to an approximated concept, e.g., by means of the boundary region size and also the approximation space size measured by an aggregation of the sizes of the approximation space components. In the process of searching for the (sub-)optimal granule, i.e., in this case a classifier, all these parameters are tuned using the minimal length principle. In searching for relevant components of approximation spaces, employing various kinds of reducts plays an important role.

8.2.2   Compound Concept Approximation

The strategies for data models inducing developed so far are often not satisfactory for approximation of compound concepts that occur in the perception process. Researchers from the different areas have recognized the necessity to work on new methods for concept approximation (see, e.g., [47, 208]). The main reason for this is that these compound concepts are, in a sense, too far from measurements which makes the searching for relevant features infeasible in a very huge space. There are several research directions aiming at overcoming this difficulty. One of them is based on the interdisciplinary research where the knowledge pertaining to perception in psychology or neuroscience is used to help to deal with compound concepts (see, e.g., [35, 90]). There is a great effort in neuroscience towards understanding the hierarchical structures of neural networks in living organisms [90]. Also mathematicians are recognizing problems of learning as the main problem of the current century [117]. These problems are closely related to complex system modeling as well. In such systems again the problem of concept approximation and its role in reasoning about perceptions is one of the challenges nowadays. One should take into account that modeling complex phenomena entails the use of local models (captured by local agents, if one would like to use the multi-agent terminology [28, 78]) that should be fused afterwards. This process involves negotiations between agents [28, 78] to resolve contradictions and conflicts in local modeling. This kind of modeling is becoming more and more important in dealing with complex real-life phenomena which we are unable to model using traditional analytical approaches. The latter approaches lead to exact models. However, the necessary assumptions used to develop them result in solutions that are too far from reality to be accepted. New methods or even a new science therefore should be developed for such modeling [40].

One of the possible approaches in developing methods for compound concept approximations can be based on the layered (hierarchical) learning [200]. Inducing concept approximation should be developed hierarchically starting from concepts that can be directly approximated using sensor measurements toward compound target concepts related to perception. This general idea can be realized using additional domain knowledge represented in natural language. For example, one can use some rules of behavior on the roads, expressed in natural language, to assess from recordings (made, e.g., by camera and other sensors) of actual traffic situations, if a particular situation is safe or not (see, e.g., [9, 10, 97]). The hierarchical learning has been also used for identification of risk patterns in medical data and extended for therapy planning (see, e.g. [6]). To deal with such problems one should develop methods for concept approximations together with methods aiming at approximation of reasoning schemes (over such concepts) expressed in natural language. The foundations of such an approach, creating a core of perception logic, are based on rough set theory [106]. The (approximate) Boolean reasoning methods can be scaled to the case of compound concept approximation.

Let us consider more examples.

Example 8.7. The prediction of behavioral patterns of a compound object evaluated over time is usually based on some historical knowledge representation used to store information about changes in relevant features or parameters. This information is usually represented as a data set and has to be collected during long-term observation of a complex dynamic system. For example, in case of road traffic, we associate the object-vehicle parameters with the readouts of different measuring devices or technical equipment placed inside the vehicle or in the outside environment (e.g., alongside the road, in a helicopter observing the situation on the road, in a traffic patrol vehicle). Many monitoring devices serve as informative sensors such as GPS, laser scanners, thermometers, range finders, digital cameras, radar, image and sound converters (see, e.g. [206]). Hence, many vehicle features serve as models of physical sensors. Here are some exemplary sensors: location, speed, current acceleration or deceleration, visibility, humidity (slipperiness) of the road. By analogy to this example, many features of compound objects are often dubbed sensors. We discuss (see also [9]) some rough set tools for perception modeling that make it possible to recognize behavioral patterns of objects and their parts changing over time. More complex behavior of compound objects or groups of compound objects can be presented in the form of *behavioral graphs*. Any behavioral graph can be interpreted as a *behavioral pattern* and can be used as a complex classifier for recognition of complex behaviors. The complete approach to the perception of behavioral patterns, based on behavioral graphs and the dynamic elimination of behavioral patterns, is presented in [9]. The tools for dynamic elimination of behavioral patterns are used for switching-off in the *system attention* procedures searching for identification of some behavioral patterns. The developed rough set tools for perception modeling are used to model networks of classifiers. Such networks make it possible to recognize behavioral patterns of objects changing over time. They are constructed using an ontology of concepts provided by experts that engage in approximate reasoning on concepts embedded in such an ontology. Experiments on data from a vehicular traffic simulator [7] are showing that the developed methods are useful in the identification of behavioral patterns.

Example 8.8. The following example concerns human computer-interfaces that allow for a dialog with experts to transfer to the system their knowledge about structurally compound objects. For pattern recognition systems [27], e.g., for Optical Character Recognition (OCR) systems it will be helpful to transfer to the system a certain knowledge about the expert's view on border line cases. The central issue in such pattern recognition systems is the construction of classifiers within vast and poorly understood search spaces, which is a very difficult task. Nonetheless, this process can be greatly enhanced with knowledge about the investigated objects provided by a human expert. We developed a framework for the transfer of such knowledge from the expert and for incorporating it into the learning process of a recognition system using methods based on rough mereology. It is also demonstrated how this knowledge acquisition can be conducted in an interactive manner, with a large dataset of handwritten digits as an example.

The next two examples are related to approximation of compound concepts in reinforcement learning and planning.

Example 8.9. In reinforcement learning [26, 64, 201], the main task is to learn the approximation of the function $Q(s, a)$, where s, a denotes a global state of the system and an action performed by an agent ag and, respectively and the real value of $Q(s, a)$ describes the reward for executing the action a in the state s. In approximation of the function $Q(s, a)$ probabilistic models are used. However, for compound real-life problems it may be hard to build such models for such a compound concept as $Q(s, a)$ [208]. We propose another approach to approximation of $Q(s, a)$ based on ontology approximation. The approach is based on the assumption that in a dialog with experts an additional knowledge can be acquired making it possible to create a ranking of values $Q(s, a)$ for different actions a in a given state s. In the explanation given by expert about possible values of $Q(s, a)$ are used concepts from a special ontology of concepts. Next, using this ontology one can follow hierarchical learning methods to learn approximations of concepts from ontology. Such concepts can have temporal character too. This means that the ranking of actions may depend not only on the actual action and the state but also on actions performed in the past and changes caused by these actions.

Example 8.10. In [6], a computer tool based on rough sets for supporting automated planning of the medical treatment is discussed. In this approach, a given patient is treated as an investigated complex dynamical system, whilst diseases of this patient (RDS, PDA, Sepsis, Ureaplasma and Respiratory Failure) are treated as compound objects changing and interacting over time. As a measure of planning success (or failure) in experiments, a special hierarchical classifier that can predict the similarity between two plans as a number between 0.0 and 1.0 is used. This classifier has been constructed on the basis of the special ontology specified by human experts and data sets. It is important to mention that besides the ontology, experts provided the exemplary data (values of attributes) for the purpose of concepts approximation from the ontology. The methods of construction such classifiers are based on approximate reasoning schemes (AR schemes, for short) and were described, e.g., in [6, 9, 10, 97]. This method was used for approximation of similarity between plans generated in automated planning and plans proposed by human experts during the realistic clinical treatment.

8.3   Modeling of Compound Granules

Methods based on information systems are crucial in modeling of compound pattern granules.

Let us first recall a generalization of information systems (see, e.g., [109]). For any attribute $a \in A$ of an information system (U, A) we consider together with the value set V_a of a a relational structure $\mathcal{R}_a$ over the universe V_a. We also consider a language $\mathcal{L}_a$ of formulas (of the same relational signature as $\mathcal{R}_a$). Such formulas interpreted over $\mathcal{R}_a$ define subsets of the Cartesian products of V_a. For example, any formula α with one free variable defines a subset $\|\alpha\|_{\mathcal{R}_a}$ of V_a. Let us observe that the relational structure $\mathcal{R}_a$ (without functions) induces a relational structure over U. Indeed, for any k-ary relation r from $\mathcal{R}_a$ one can define a k-ary relation $g_a \subseteq V_a^k$ by $(x_1, \ldots, x_k) \in g_a$ if and only if $(a(x_1), \ldots, a(x_k)) \in r$ for any $(x_1, \ldots, x_k) \in U^k$. Hence, one can consider any formula from $\mathcal{L}_a$ as a constructive method of defining a subset of the universe U with a structure induced by $\mathcal{R}_a$. Any such a structure is a new information granule. On the next level of hierarchical modeling, i.e., in constructing new information systems we use such structures as objects and attributes are properties of such structures. Next, one can consider the similarity between new constructed objects and then their similarity neighborhoods will correspond to clusters of relational structures. This process is usually more complex. This is because instead of relational structure $\mathcal{R}_a$ we usually consider a fusion of relational structures corresponding to some attributes from A. The fusion makes it possible to describe constraints that should hold between parts obtained by composition from less compound parts. Examples of relational structures can be defined by indiscernibility, similarity, intervals obtained in discretization or symbolic value grouping, preference or spatio-temporal relations (see, e.g., [69, 145]). One can see that parameters to be tuned in searching for relevant (for target concept approximation) patterns over new information systems are, among others, relational structures over value sets, the language of formulas defining parts, and constraints.

The main basic steps in hierarchical modeling are the following:

1. the structures of granules on a higher level are constructed from structures of granules on the lower level;

2. a language for expressing properties of structures on a higher level is selected;
3. some formulas (features) of the structures on a higher level are selected as relevant for pattern granule construction;
4. indiscernibility (or tolerance) classes defined by a new constructed information system are used as pattern granules on the higher level.

In the following sections, we discuss in more detail some issues related to the outlined modeling.

8.3.1   Constrained Sums of Granules

One of the main task in granular computing is to develop calculi of granules [152], [154]. Information systems used in rough set theory are a particular kind of granules. In the section we study operations on such granules, basic for reasoning in distributed systems of granules. The operations are called constrained sums. They are developed by interpreting infomorphisms between classifications [4]. In [157] we have shown that classifications [4] and information systems [106] are, in a sense, equivalent. Operations, called constrained sums, seem to be very important in searching for patterns in data mining (e.g., in spatio-temporal reasoning) or in more general sense in generating relevant granules for approximate reasoning using calculi on granules [157].

First we recall the definition of infomorphism for two information systems [157]. The infomorphisms for classifications are introduced and studied in [4].

For all formulas $\alpha \in L_{IS}$ and for all objects $x \in U$ we will denote $x \models_{IS} \alpha$ if and only if $x \in Sem_{IS}(\alpha)$.

Definition 8.11. *[157] If $IS_1 = (U_1, A_1)$ and $IS_2 = (U_2, A_2)$ are information systems then an infomorphism between IS_1 and IS_2 is a pair $(f^\wedge, f^\vee)$ of functions $f^\wedge : L_{IS_1} \to L_{IS_2}$, $f^\vee : U_2 \to U_1$, satisfying the following equivalence*

$$f^\vee(x) \models_{IS_1} \alpha \text{ if and only if } x \models_{IS_2} f^\wedge(\alpha) \tag{8.8}$$

for all objects $x \in U_2$ and for all formulas $\alpha \in L_{IS_1}$.

The infomorphism will be denoted shortly by

$$(f^\wedge, f^\vee) : IS_1 \rightleftarrows IS_2.$$

8.3.2   Sum of Information Systems

In this section we discuss a sum of two information systems.

Definition 8.12. *Let $IS_1 = (U_1, A_1)$ and $IS_2 = (U_2, A_2)$ be information systems. These information systems can be combined into a single information system, denoted by $+(IS_1, IS_2)$, with the following properties:*

- *The objects of $+(IS_1, IS_2)$ consist of pairs (x_1, x_2) of objects from IS_1 and IS_2 i.e. $U = U_1 \times U_2$*

- *The attributes of $+(IS_1, IS_2)$ consist of the attributes of IS_1 and IS_2, except that if there are any attributes in common, then we make distinct copies, so as not to confuse them.*

Proposition 8.13. *There are infomorphisms $(f_k^\wedge, f_k^\vee) : IS_k \rightleftarrows +(IS_1, IS_2)$ for $k = 1, 2$ defined as follows:*

- *$f_k^\wedge(\alpha) = \alpha_{IS_k}$ (the IS_k-copy of α) for each $\alpha \in L_{IS_k}$*
- *for each pair $(x_1, x_2) \in U$, $f_k^\vee((x_1, x_2)) = x_k$*

Given any information system IS_3 and infomorphisms $(f_{k,3}^\wedge, f_{k,3}^\vee) : IS_k \rightleftarrows IS_3$, there is a unique infomorphism $(f_{1+2,3}^\wedge, f_{1+2,3}^\vee) : +(IS_1, IS_2) \rightleftarrows IS_3$ such that in Figure 8.2 one can go either way around the triangles and get the same result.

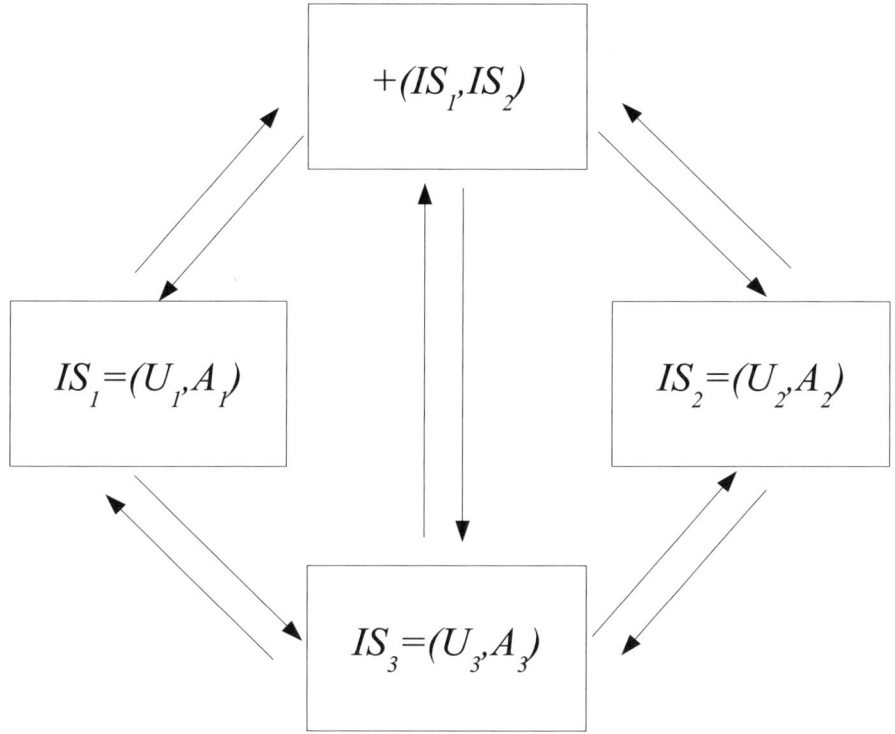

Fig. 8.2. Sum of Information Systems $IS_1 = (U_1, A_1)$ and $IS_2 = (U_2, A_2)$

Example 8.14. Let us consider a diagnostic agent testing failures of the space robotic arm. Such an agent should observe the arm and detect a failure if, e.g., some of its parts are in abnormal relative position. Let's assume, in our simple example, that projections of some parts on a plane are observed and a failure is detected if the projections of some triangular or rectangular parts are in some relation, e.g., the triangle is not included sufficiently inside the rectangle. Hence,

Table 8.1. Information System $IS_{rectangle}$ with Uncertainty Functions

$U_{rectagle}$	a	b	$I_a(\cdot)$	$I_b(\cdot)$	$I_{A_1}(\cdot)$
x_1	165	yes	$\{x_1, x_3, x_5, x_6\}$	$\{x_1, x_3\}$	$\{x_1, x_3\}$
x_2	175	no	$\{x_2, x_4, x_6\}$	$\{x_2, x_4, x_5, x_6\}$	$\{x_2, x_4, x_6\}$
x_3	160	yes	$\{x_1, x_3, x_5\}$	$\{x_1, x_3\}$	$\{x_1, x_3\}$
x_4	180	no	$\{x_2, x_4\}$	$\{x_2, x_4, x_5, x_6\}$	$\{x_2, x_4\}$
x_5	160	no	$\{x_1, x_3, x_5\}$	$\{x_2, x_4, x_5, x_6\}$	$\{x_5\}$
x_6	170	no	$\{x_1, x_2, x_6\}$	$\{x_2, x_4, x_5, x_6\}$	$\{x_2, x_6\}$

Table 8.2. Information System $IS_{triangle}$ with Uncertainty Function I_{A_2}

$U_{triangle}$	c	$I_{A_2}(\cdot)$
y_1	t_1	$\{y_1, y_3\}$
y_2	t_2	$\{y_2\}$
y_3	t_1	$\{y_1, y_3\}$

any considered object consists of parts: a triangle and a rectangle. Objects are perceived by some attributes expressing properties of parts and a relation (constraint) between them.

First, we construct an information system, called the sum of given information systems. Such system represents objects composed from parts without any constraint. It means that we consider as the universe of objects the Cartesian product of the universes of parts (Tables 8.1-8.3).

Let us consider three information systems

$$IS_{rectangle} = (U_{rectangle}, A_{rectangle}), IS_{triangle} = (U_{triangle}, A_{triangle})$$

and $+(IS_{rectangle}, IS_{triangle}) = (U_{rectangle} \times U_{triangle}, \{(a,1), (b,1), (c,2)\})$ presented in Table 8.1, Table 8.2 and Table 8.3, respectively. Let $U_{rectangle}$ be a set of rectangles and $A_{rectangle} = \{a, b\}$, $V_a = [0, 300]$ and $V_b = \{yes, no\}$, where the value of a means a length in millimeters of horizontal side of rectangle and for any object $x \in U_{rectangle}$ $b(x) = yes$ if and only if x is a square.

Let $U_{triangle}$ be a set of triangles and $A_{triangle} = \{c\}$ and $V_c = \{t_1, t_2\}$, where $c(x) = t_1$ if and only if x is an acute-angled triangle and $c(x) = t_2$ if and only if x is a right-angled triangle.

We assume all values of attributes are made on a given projection plane. The results of measurements are represented in information systems. Tables 8.1-8.2 include only illustrative examples of the results of such measurements. We define uncertainty functions as follows

$$y \in I_a(x) \text{ if and only if } |a(x) - a(y)| \leq 5$$

$$y \in I_b(x) \text{ if and only if } b(x) = b(y)$$

$$y \in I_{A_1}(x) \text{ if and only if } (y \in I_a(x) \text{ and } y \in I_b(x))$$

We assume that $(a, 1)((x_i, y_j)) = a(x_i)$, $(b, 1)((x_i, y_j)) = b(x_i)$ and $(c, 2)((x_i, y_j)) = c(y_j)$, where $i = 1, \ldots, 6$ and $j = 1, 2$.

Table 8.3. An Information System $+(IS_{rectangle}, IS_{triangle})$ with Uncertainty Function I_{A_1, A_2}

$U_{rectangle} \times U_{triangle}$	$(a,1)$	$(b,1)$	$(c,2)$	$I_{A_1, A_2}((\cdot, \cdot))$
(x_1, y_1)	165	yes	t_1	$\{x_1, x_3\} \times \{y_1, y_3\}$
(x_1, y_2)	165	yes	t_2	$\{x_1, x_3\} \times \{y_2\}$
(x_1, y_3)	165	yes	t_1	$\{x_1, x_3\} \times \{y_1, y_3\}$
(x_2, y_1)	175	no	t_1	$\{x_2, x_4, x_6\} \times \{y_1, y_3\}$
(x_2, y_2)	175	no	t_2	$\{x_2, x_4, x_6\} \times \{y_2\}$
(x_2, y_3)	175	no	t_1	$\{x_2, x_4, x_6\} \times \{y_1, y_3\}$
(x_3, y_1)	160	yes	t_1	$\{x_1, x_3\} \times \{y_1, y_3\}$
(x_3, y_2)	160	yes	t_2	$\{x_1, x_3\} \times \{y_2\}$
(x_3, y_3)	160	yes	t_1	$\{x_1, x_3\} \times \{y_1, y_3\}$
(x_4, y_1)	180	no	t_1	$\{x_2, x_4\} \times \{y_1, y_3\}$
(x_4, y_2)	180	no	t_2	$\{x_2, x_4\} \times \{y_2\}$
(x_4, y_3)	180	no	t_1	$\{x_2, x_4\} \times \{y_1, y_3\}$
(x_5, y_1)	160	no	t_1	$\{x_5\} \times \{y_1, y_3\}$
(x_5, y_2)	160	no	t_2	$\{x_5\} \times \{y_2\}$
(x_5, y_3)	160	no	t_1	$\{x_5\} \times \{y_1, y_3\}$
(x_6, y_1)	170	no	t_1	$\{x_2, x_6\} \times \{y_1, y_3\}$
(x_6, y_2)	170	no	t_2	$\{x_2, x_6\} \times \{y_2\}$
(x_6, y_3)	170	no	t_1	$\{x_2, x_6\} \times \{y_1, y_3\}$

8.3.3   Sum of Approximation Spaces

In this section we present a simple construction of approximation space for the sum of given approximation spaces.

Let $AS_{\#_k} = (U_k, I_{\#_k}, \nu_{SRI})$ be an approximation space for information system IS_k, where $k = 1, 2$.

We define an approximation space $+(AS_{\#_1}, AS_{\#_2})$ for information system $+(IS_1, IS_2)$ as follows:

1. the universe is equal to $U_1 \times U_2$;
2. $I_{\#_1, \#_2}((x_1, x_2)) = I_{\#_1}(x_1) \times I_{\#_2}(x_2)$;
3. the inclusion function ν_{SRI} in $+(AS_{\#_1}, AS_{\#_2})$ is the standard rough inclusion function.

We have the following property:

Proposition 8.15. *For any subsets $X \subseteq U_1$ and $Y \subseteq U_2$ we obtain*

$$LOW(+(AS_{\#_1}, AS_{\#_2}), X \times Y) =$$
$$LOW(AS_{\#_1}, X) \times LOW(AS_{\#_2}, Y) \tag{8.9}$$
$$UPP(+(AS_{\#_1}, AS_{\#_2}), X \times Y) =$$
$$UPP(AS_{\#_1}, X) \times UPP(AS_{\#_2}, Y). \tag{8.10}$$

Proof. We assume that $I_{\#_1} : U_1 \to P(U_1)$, $I_{\#_2} : U_2 \to P(U_2)$ and $I_{\#_1, \#_2} : U_1 \times U_2 \to P(U_1 \times U_2)$. Let $x_1 \in U_1$ and $x_2 \in U_2$. We have $I_{\#_1, \#_2}((x_1, x_2)) \subseteq X \times Y$

if and only if $I_{\#_1}(x_1) \subseteq X$ and $I_{\#_2}(x_2) \subseteq Y$. Moreover, $I_{\#_1,\#_2}((x_1,x_2)) \cap (X \times Y) \neq \emptyset$ if and only if $I_{\#_1}(x_1) \cap X \neq \emptyset$ and $I_{\#_2}(x_2) \cap Y \neq \emptyset$.

Example 8.16. For information system $IS_{rectangle}$ we define an approximation space $AS_{A_1} = (U_{rectangle}, I_{A_1}, \nu_{SRI})$ such that

$y \in I_a^5(x)$ if and only if $|a(x) - a(y)| \leq 5$. This means that rectangles x and y are similar with respect to the length of horizontal sides if and only if the difference in lengths is not greater than 5 millimeters.

Let $y \in I_b(x)$ if and only if $b(x) = b(y)$ and

$y \in I_{A_1}(x)$ if and only if $\forall_{c \in A_1} y \in I_c(x)$

Thus, we obtain uncertainty functions represented in the last three columns of Table 8.1.

For the information system $IS_{triangle}$ we define an approximation space as follows:

$y \in I_{A_2}(x)$ if and only if $c(x) = c(y)$ (see the last column of Table 8.2).

For $+(IS_{rectangle}, IS_{triangle})$ we obtain $I_{A_1,A_2}((x,y)) = I_{A_1}(x) \times I_{A_2}(y)$ (see the last column of Table 8.3).

8.3.4 Sum with Constraints of Information Systems

In this section we consider a new operation on information systems often used in searching, e.g., for relevant patterns. We start from the definition in which the constraints are given explicitly.

Definition 8.17. *Let $IS_i = (U_i, A_i)$ for $i = 1, \dots, k$ be information systems and let R be a k-ary constraint relation in $U_1 \times \dots \times U_k$, i.e., $R \subseteq U_1 \times \dots \times U_k$. These information systems can be combined into a single information system relatively to R, denoted by $+_R(IS_1, \dots, IS_k)$, with the following properties:*

- *The objects of $+_R(IS_1, \dots, IS_k)$ consist of k-tuples $(x_1, \dots, x_k)$ of objects from R, i.e., all objects from $U_1 \times \dots \times U_k$ satisfying the constraint R.*
- *The attributes of $+_R(IS_1, \dots, IS_k)$ consist of the attributes from the sets $A_1, \dots, A_k$, except that if there are any attributes in common, then we make distinct copies, so as not to confuse them.*

Usually the constraints are defined by conditions expressed by Boolean combination of descriptors of attributes. It means that the constraints are built from expressions $a = v$, where a is an attribute and v is its value, using propositional connectives $\wedge, \vee, \neg$. Observe, that in constraint definition we use not only attributes of parts (i.e., from information systems $IS_1, \dots, IS_k$) but also some other attributes specifying relation between parts. In our example (see Table 8.4), the constraint R_1 is defined as follows: *the triangle is sufficiently included in the rectangle.* Any row of this table represents an object (x_i, y_j) composed of the triangle y_j included sufficiently into the rectangle x_i.

Let us also note that constraints are defined using primitive (measurable) attributes different than those from information systems describing parts. This makes the sum with constraint operation different from theta join. On the other

Table 8.4. Information System $+_{R_1}(IS_{rectangle}, IS_{triangle})$

$(U_{rectangle} \times U_{triangle}) \cap R_1$	a'	b'	c'
(x_1, y_1)	165	yes	t_1
(x_1, y_2)	165	yes	t_2
(x_2, y_1)	175	no	t_1
(x_2, y_2)	175	no	t_2
(x_3, y_1)	160	yes	t_1
(x_3, y_2)	160	yes	t_2
(x_4, y_1)	180	no	t_1
(x_4, y_2)	180	no	t_2
(x_5, y_1)	160	no	t_1
(x_5, y_2)	160	no	t_2
(x_6, y_1)	170	no	t_1
(x_6, y_2)	170	no	t_2

hand one can consider that the constraints are defined in two steps. In the first step we extend the attributes for parts and in the second step we define the constraints using some relations on these new attributes.

Let us observe that the information system $+_R(IS_1, \ldots, IS_k)$ can be also described using an extension of the sum $+(IS_1, \ldots, IS_k)$ by adding a new binary attribute $a_R : U_1 \times \ldots \times U_k \rightarrow \{0, 1\}$ that is the characteristic function of the relation R and by taking a subsystem of the received system consisting of all objects having value one for this new attribute.

The constraints used to define the sum (with constraints) can be often specified by information systems. The objects of such systems are tuples consisting of objects of information systems that are arguments of the sum. The attributes describe relations between elements of tuples. One of the attribute is a characteristic function of the constraint relation (restricted to the universe of the information system). In this way we obtain a decision system with the decision attribute defined by the characteristic function of the constraint and conditional attributes are the remaining attributes of this system. From such decision table one can induce classifier for the constraint relation. Next, the classifier can be used to select tuples in the construction of the sum with constraints.

Example 8.18. Let us consider three information systems

$$IS_{rectangle} = (U_{rectangle}, A_{rectangle}), IS_{triangle} = (U_{triangle}, A_{triangle}),$$

$$+_{R_1}(IS_{rectangle}, IS_{triangle}),$$

presented in Table 8.1, Table 8.2 and Table 8.4, respectively. We assume that $R_1 = \{(x_i, y_j) \in U_{rectangle} \times U_{triangle} : i = 1, \ldots, 6 \quad j = 1, 2\}$. We also assume that $a'((x_i, y_j)) = a(x_i)$, $b'((x_i, y_j)) = b(x_i)$ and $c'((x_i, y_j)) = c(y_j)$, where $i = 1, \ldots, 6$ and $j = 1, 2$.

The above examples are illustrating an idea of specifying constraints by examples. Table 8.4 can be used to construct a decision table partially specifying characteristic functions of the constraint. Such a decision table should be extended by adding relevant attributes related to the object parts, which allows to induce the high quality classifiers for the constraint relation. The classifier can then be used to filter composed pairs of objects that satisfy the constraint. This is an important construction because the constraint specification usually cannot be defined directly in terms of measurable attributes. It can be specified, e.g., in natural language. This is the reason that the process of inducing of the relevant classifiers for constraints can require hierarchical classifier construction [104].

The constructed constraint sum of information systems can contain some incorrect objects. This is due to improper filtering of objects by the constraint classifier induced from data (with accuracy usually less than 100%). One should take this issue into account in constructing nets of information systems.

8.3.5   Constraint Sum of Approximation Spaces

Let $AS_{\#_i} = (U_i, I_{\#_i}, \nu_{SRI})$ be an approximation space for information system IS_i, where $i = 1, \ldots, k$ and let $R \subseteq U_1 \times \ldots \times U_k$ be a constraint relation. We define an approximation space $+_R(AS_{\#_1}, \ldots, AS_{\#_k})$ for $+_R(IS_1, \ldots, IS_k)$ as follows:

1. the universe is equal to R;
2. $I_{\#_1, \ldots, \#_k}((x_1, \ldots, x_k)) = (I_{\#_1}(x_1) \times \ldots \times I_{\#_k}(x_k)) \cap R$;
3. the inclusion relation ν_{SRI} in $+_R(AS_{\#_1}, \ldots, AS_{\#_k})$ is the standard inclusion function.

We have the following properties of approximations:

Proposition 8.19

$$LOW(+_R(AS_{\#_1}, \ldots, AS_{\#_k}), X_1 \times \ldots \times X_k) =$$
$$R \cap (LOW(AS_{\#_1}, X_1) \times \ldots \times LOW(AS_{\#_k}, X_k)) \tag{8.11}$$
$$UPP(+_R(AS_{\#_1}, \ldots, AS_{\#_k}), X_1 \times \ldots \times X_k) =$$
$$R \cap (UPP(AS_{\#_1}, X_1) \times \ldots \times UPP(AS_{\#_k}, X_k)). \tag{8.12}$$

8.4   Rough–Fuzzy Granules

In this section, we discuss rough-fuzzy granules. Such granules are important for many applications. To explain the main idea behind such granules let us consider the problem of construction of a classifier for a vague concept specified by a sample of positive and negative examples. Quite often, the induced boundary region of the concept can be too large and later the information that a given object falls into the boundary region may not be that meaningful in applications. In such cases, one can try to distinguish different parts in the boundary region representing different *shades* of the concept. Next, with these parts being treated as new concepts, their approximations are constructed. For applications, it is very important to have linearly ordered parts, called layers. The boundary regions of layers should satisfy the constraint that each of them has non-empty intersection with two neighboring layers only. Next, approximations of layers are extended from a given sample to the whole space of objects. The induced membership functions for the parts can be treated as rough-fuzzy membership functions of linguistic variables [221] corresponding to parts (e.g., *low, medium, high*). In this way, we obtain a family of classifiers as an approximation of a given concept. We call such a family a rough-fuzzy granule.

Below we present a more formal description of the above idea.

Let $DT = (U, A, d)$ be a decision table where the decision d is the fuzzy membership function ν restriction to the objects from U. Consider reals $0 < c_1 < \ldots < c_k$ where $c_i \in (0, 1]$ for $i = 1, \ldots, k$. Any c_i defines c_i-cut by $X_i = \{x \in U : \nu(x) \geq c_i\}$. Assume that $X_0 = U$ and $X_{k+1} = X_{k+2} = \emptyset$. A *rough-fuzzy granule* (*rf-granule*, for short) corresponding to $(DT, c_1, \ldots, c_k)$ is any granule $g = (g_0, \ldots, g_k)$ such that for some $B \subseteq A$,

$$Sem_B(g_i) = [LOW(AS_B, (X_i - X_{i+1})), UPP(AS_B, (X_i - X_{i+1}))], \quad (8.13)$$
$$\text{for } i = 0, \ldots, k, \text{ and}$$
$$UPP(AS_B, (X_i - X_{i+1})) \subseteq (X_{i-1} - X_{i+2}), \text{ for } i = 1, \ldots, k,$$

where $Sem_B(g_i)$ denotes the semantics of g_i.

Any function $\nu^* : U \to [0, 1]$ satisfying the conditions

$$\nu^*(x) = 0, \text{ for } x \in U - UPP(AS_B, X_1), \quad (8.14)$$
$$\nu^*(x) = 1, \text{ for } x \in LOW(AS_B, X_k),$$
$$\nu^*(x) = c_{i-1}, \text{ for } x \in LOW(AS_B, (X_{i-1} - X_i)), \text{ and } i = 2, \ldots, k-1,$$
$$c_{i-1} < \nu^*(x) < c_i, \text{ for } x \in (UPP(AS_B, X_i) - LOW(AS_B, X_i)),$$
$$\text{where } i = 1, \ldots, k, \text{ and } c_0 = 0,$$

is called a B-approximation of ν.

For applications, it is necessary to develop heuristics searching for relevant attributes and parts as well as their approximations. The constructed rough-fuzzy granules are used, e.g., in approximation of other concepts.

8.5   Conclusions

We have discussed some issues for intelligent systems based on granular computing. The most important are applications of granules for mining complex knowledge from complex data (one of challenging problems in data mining research [218]).

The approach can be extended to adaptive approximation of concepts in multi-agent systems, e.g., for control of complex adaptive systems (this problem is related to distributed data mining and mining multi-agent data [218]).

In our opinion rough–granular computing is a good foundation for "developing a unifying theory of data mining" [218].

Part IV

Conclusions, Bibliography and Further Readings

9 Concluding Remarks

In this book we have outlined a methodology for knowledge discovery and data mining by means of rough–granular computing. Several research directions are related to rough–granular computing. We enclose a list of such directions together with examples of problems.

1. **Developing foundations for information granule systems.** Certainly, still more work is needed to develop solid foundations for synthesizing and analyzing information granule systems. In particular, methods for constructing hierarchical information granule systems and methods for representing such systems should be developed.
2. **Algorithmic methods for inducing parameterized approximation spaces from data and background knowledge.** Some methods have already been reported, such as the discovery of approximation space based on discretization of attributes and methods based on distance functions. However, these are only initial steps toward algorithmic methods for inducing parameterized approximation spaces from data and background knowledge.
3. **Algorithmic methods for synthesizing of approximate reasoning schemes.** It was observed that problems of negotiation and conflict resolution are of great importance in synthesizing of approximate reasoning schemes. The problem arises, e.g., when we are searching in a given set of agents for a granule sufficiently included or close to a given one. These agents, often working with different systems of information granules, can derive different granules, and their fusion will be necessary to obtain the relevant output granule. In the fusion process, negotiations and conflict resolutions are necessary. Much more work should be done in this direction by using the existing results on negotiations and conflict resolution. In particular, Boolean reasoning methods seem to be promising.
4. **Fusion methods in rough–granular computing.** A basic problem is fusion of the inputs (information) derived from information granules. This fusion makes it possible to contribute to the construction of new granules.

J. Stepaniuk: Rough - Gran. Comput. in Knowl. Dis. & Data Min., SCI 152, pp. 135–136, 2008.
springerlink.com

5. **Discovery of multiagent systems relevant to given problems.** Quite often, agents and communication methods among them are not given a priori with the problem specification, and the challenge is to develop methods for discovering multiagent system structures relevant to given problems, in particular, methods for discovering relevant communication protocols.

References

1. Abidi, S.S.R., Hoe, K.M., Goh, A.: Analyzing Data Clusters: A Rough Sets Approach to Extract Cluster-Defining Symbolic Rules. In: Hoffmann, F., Adams, N., Fisher, D., Guimarães, G., Hand, D.J. (eds.) IDA 2001. LNCS, vol. 2189, pp. 248–257. Springer, Heidelberg (2001)
2. Agrawal, R., Mannila, H., Srikant, R., Toivonen, H., Verkano, A.: Fast Discovery of Association Rules. In: Fayyad, U.M., Piatetsky-Shapiro, G., Smyth, P., Uthurusamy, R. (eds.) Advances in Knowledge Discovery and Data Mining, pp. 307–328. The AAAI Press/The MIT Press (1996)
3. Bargiela, A., Pedrycz, W.: Granular Computing: An Introduction. Kluwer Academic Publishers, Dordrecht (2003)
4. Barwise, J., Seligman, J.: Information Flow: The Logic of Distributed Systems. Tracts in Theoretical Computer Science, vol. 44. Cambridge University Press, Cambridge (1997)
5. Bazan, J.G.: A Comparison of Dynamic and Non–Dynamic Rough Set Methods for Extracting Laws from Decision Tables. In: Polkowski, L., Skowron, A. (eds.) Rough Sets in Knowledge Discovery 1. Methodology and Applications, pp. 321–365. Physica–Verlag, Heidelberg (1998)
6. Bazan, J., Kruczek, P., Bazan-Socha, S., Skowron, A., Pietrzyk, J.J.: Risk Pattern Identification in the Treatment of Infants with respiratory failure through rough set modeling. In: Proceedings of IPMU 2006, Paris, France, July 2-7, 2006, pp. 2650–2657. Éditions E.D.K., Paris (2006)
7. Bazan, J., Kruczek, P., Bazan-Socha, S., Skowron, A., Pietrzyk, J.J.: Rough Set Approach to Behavioral Pattern Identification. Fundamenta Informaticae 75(1-4), 27–47 (2007)
8. Bazan, J., Nguyen, H.S., Nguyen, S.H., Skowron, A.: Rough set methods in approximation of hierarchical concepts. In: Tsumoto, S., Słowiński, R., Komorowski, J., Grzymała-Busse, J.W. (eds.) RSCTC 2004. LNCS (LNAI), vol. 3066, pp. 346–355. Springer, Heidelberg (2004)
9. Bazan, J., Peters, J.F., Skowron, A.: Behavioral Pattern Identification Through Rough Set Modelling. In: Ślęzak, D., Yao, J., Peters, J.F., Ziarko, W., Hu, X. (eds.) RSFDGrC 2005. LNCS (LNAI), vol. 3642, pp. 688–697. Springer, Heidelberg (2005)

10. Bazan, J., Skowron, A.: Classifiers based on approximate reasoning schemes. In: Dunin-Keplicz, B., Jankowski, A., Skowron, A., Szczuka, M. (eds.) Monitoring, Security, and Rescue Tasks in Multiagent Systems MSRAS. Advances in Soft Computing, pp. 191–202. Springer, Heidelberg (2005)
11. Bazan, J., Nguyen, H.S., Nguyen, T.T., Skowron, A., Stepaniuk, J.: Some Logic and Rough Set Applications for Classifying Objects. Institute of Computer Science, Warsaw University of Technology, ICS Research Report, 38/94 (1994)
12. Bazan, J., Nguyen, H.S., Nguyen, T.T., Skowron, A., Stepaniuk, J.: Application of Modal Logics and Rough Sets for Classifying Objects. In: De Glas, M., Pawlak, Z. (eds.) Proceedings of the Second World Conference on Fundamentals of Artificial Intelligence (WOCFAI 1995), Paris, July 3-7, pp. 15–26. Angkor, Paris (1995)
13. Bazan, J., Nguyen, H.S., Nguyen, T.T., Skowron, A., Stepaniuk, J.: Synthesis of Decision Rules for Object Classification. In: Orlowska, E. (ed.) Incomplete Information: Rough Set Analysis, pp. 23–57. Physica-Verlag, Heidelberg (1998)
14. Bazan, J., Skowron, A., Swiniarski, R.: Rough sets and vague concept approximation: From sample approximation to adaptive learning. In: Peters, J.F., Skowron, A. (eds.) Transactions on Rough Sets V. LNCS, vol. 4100, pp. 39–62. Springer, Heidelberg (2006)
15. Bazan, J., Szczuka, M.: The Rough Set Exploration System. In: Peters, J.F., Skowron, A. (eds.) Transactions on Rough Sets III. LNCS, vol. 3400, pp. 37–56. Springer, Heidelberg (2005)
16. Bonchi, F., Boulicaut, J.-F. (eds.): KDID 2005. LNCS, vol. 3933. Springer, Heidelberg (2006)
17. Brazdil, P., Torgo, L.: Knowledge Acquisition via Knowledge Integration, Current Trends in Knowledge Acqusition. IOS Press, Amsterdam (1990)
18. Breiman, L.: Statistical Modeling: The Two Cultures. Statistical Science 16(3), 199–231 (2001)
19. Bruha, I.: Quality of Decision Rules: Definitions and Classification Schemes for Multiple Rules. In: Nakhaeizadeh, G., Taylor, C.C. (eds.) Machine Learning and Statistics, The Interface, pp. 107–131. John Wiley and Sons, Chichester (1997)
20. Cattaneo, G.: Abstract Approximation Spaces for Rough Theories. In: Polkowski, L., Skowron, A. (eds.) Rough Sets in Knowledge Discovery 1. Methodology and Applications, pp. 59–98. Physica–Verlag, Heidelberg (1998)
21. Cios, K.J., Pedrycz, W., Świniarski, R.W., Kurgan, L.A.: Data Mining A Knowledge Discovery Approach. Springer, Heidelberg (2007)
22. Cox, T.F., Cox, M.A.: Multidimensional Scaling, Monographs on Statistics and Applied Probability. Chapman-Hall, London (1994)
23. Czyżewski, A., Kostek, B.: Rough Set-Based Filtration of Sound Applicable to Hearing Prostheses. In: Tsumoto, S., Kobayashi, S., Yokomori, T., Tanaka, H. (eds.) Proceedings of the Fourth International Workshop on Rough Sets, Fuzzy Sets and Machine Discovery (RSFD 1996), Tokyo, November 6-8, pp. 168–175 (1996)
24. Czyżewski, A., Królikowski, R., Skórka, P.: Automatic Detection of Speech Disorders. In: Proceedings of the Fourth European Congress on Intelligent Techniques and Soft Computing, Aachen, Germany, September 2-5, vol. 1, pp. 183–187 (1996)
25. Dasarathy, B.V. (ed.): Nearest Neighbor Pattern Classification Techniques. IEEE Computer Society Press, Los Alamitos (1991)
26. Dietterich, T.G.: Hierarchical reinforcement learning with the MAXQ value function decomposition. Artificial Intelligence 13(5), 227–303 (2000)
27. Duda, R., Hart, P., Stork, R.: Pattern Classification. John Wiley & Sons, New York (2002)

28. Dunin-Kęplicz, B., Jankowski, A., Skowron, A., Szczuka, M. (eds.): Monitoring, Security, and Rescue Tasks in Multiagent Systems (MSRAS 2004). Advances in Soft Computing. Springer, Heidelberg (2005)

29. Düntsch, I.: Rough Sets and Algebras of Relations. In: Orlowska, E. (ed.) Incomplete Information: Rough Set Analysis, pp. 95–108. Physica-Verlag, Heidelberg (1998)

30. Dzeroski, S., Lavrac, N. (eds.): Relational Data Mining. Springer, Berlin (2001)

31. El-Mouadib, F.A., Koronacki, J., Żytkow, J.M.: Taxonomy Formation by Approximate Equivalence Relations. In: Żytkow, J.M., Rauch, J. (eds.) PKDD 1999. LNCS (LNAI), vol. 1704, pp. 71–79. Springer, Heidelberg (1999)

32. Ester, M., et al.: A Density-Based Algorithm for Discovering Clusters in Large Spatial Databases with Noise. In: Proceedings of 2nd International Conference Knowledge Discovery and Data Mining, pp. 226–231. AAAI-Press, Portland (1996)

33. Fahle, M., Poggio, T. (eds.): Perceptual Learning. MIT Press, Cambridge (2002)

34. Fayyad, U.M., Piatetsky-Shapiro, G., Smyth, P., Uthurusamy, R. (eds.): Advances in Knowledge Discovery and Data Mining. AAAI Press/The MIT Press (1996)

35. Forbus, K.D., Hinrisch, T.R.: Engines of the brain: The computational instruction set of human cognition. AI Magazine 27, 15–31 (2006)

36. Fraley, C., Raftery, A.E.: How Many Clusters? Which Clustering Method? Answers via Model-Based Cluster Analysis, Technical Report No.329, University of Washington, USA (1998)

37. Frege, G.: Grundlagen der Arithmetik 2. Verlag von Herman Pohle, Jena (1893)

38. Funakoshi, K., Ho, T.B.: Information Retrieval by Rough Tolerance Relation. In: Tsumoto, S., Kobayashi, S., Yokomori, T., Tanaka, H. (eds.) Proceedings of the Fourth International Workshop on Rough Sets, Fuzzy Sets and Machine Discovery (RSFD 1996), Tokyo, November 6-8, pp. 31–35 (1996)

39. Funakoshi, K., Ho, T.B.: A Rough Set Approach to Information Retrieval. In: Polkowski, L., Skowron, A. (eds.) Rough Sets in Knowledge Discovery 2. Applications, Case Studies and Software Systems, pp. 166–177. Physica-Verlag, Heidelberg (1998)

40. Gell-Mann, M.: The Quark and the Jaguar - Adventures in the Simple and the Complex. Little, Brown and Co., London (1994)

41. Gemello, R., Mana, F.: An Integrated Characterization and Discrimination Scheme to Improve Learning Efficiency in Large Data Sets. In: Proceedings of the Eleventh International Joint Conference on Artificial Intelligence, Detroit MI, August 20-25, pp. 719–724 (1989)

42. Gomolinska, A.: A Comparison of Pawlak's and Skowron-Stepaniuk's Approximation of Concepts. In: Peters, J.F., Skowron, A., Düntsch, I., Grzymała-Busse, J.W., Orłowska, E., Polkowski, L. (eds.) Transactions on Rough Sets VI. LNCS, vol. 4374, pp. 64–82. Springer, Heidelberg (2007)

43. Greco, S., Matarazzo, B., Słowiński, R.: Rough Approximation of a Preference Relation in a Pairwise Comparison Table. In: Polkowski, L., Skowron, A. (eds.) Rough Sets in Knowledge Discovery 2. Applications, Case Studies and Software Systems, pp. 13–36. Physica-Verlag, Heidelberg (1998)

44. Grzymała-Busse, J.W.: A New Version of the Rule Induction System LERS. Fundamenta Informaticae 31, 27–39 (1997)

45. Grzymała-Busse, J.W.: Applications of the Rule Induction System LERS. In: Polkowski, L., Skowron, A. (eds.) Rough Sets in Knowledge Discovery 1. Methodology and Applications, pp. 366–375. Physica–Verlag, Heidelberg (1998)

46. Halkidi, M., Batistakis, Y., Vazirgiannis, M.: On clustering validation techniques. Journal of Intelligent Information Systems 17(2/3), 107–145 (2001)
47. Hastie, T., Tibshirani, R., Friedman, J.: The Elements of Statistical Learning. Springer, Heidelberg (2001)
48. Hirano, S., Tsumoto, S.: On Constructing Clusters from Non-Euclidean Dissimilarity Matrix by Using Rough Clustering. In: Washio, T., Sakurai, A., Nakajima, K., Takeda, H., Tojo, S., Yokoo, M. (eds.) JSAI Workshop 2006. LNCS (LNAI), vol. 4012, pp. 5–16. Springer, Heidelberg (2006)
49. Hobbs, J.R.: Granularity. In: Proceedings of Ninth International Joint Conference on Artificial Intelligence, Los Angeles, California, pp. 432–435 (August 1985); Weld, D.S., de Kleer, J. (eds.) Readings in Qualitative Reasoning about Physical Systems, pp. 542–545. Morgan Kaufmann Publishers, Inc., San Mateo, California (1989)
50. Hobbs, J.R.: Half Orders of Magnitude. In: KR 2000 Workshop on Semantic Approximation, Granularity, and Vagueness, Breckenridge, Colorado (April 2000)
51. Hobbs, J.R., Kreinovich, V.: Optimal Choice of Granularity in Commonsense Estimation: Why Half Orders of Magnitude. In: Proceedings of Joint 9th IFSA World Congress and 20th NAFIPS International Conference, Vacouver, British Columbia, July 2001, pp. 1343–1348 (2001)
52. Holte, R.C.: Very Simple Classification Rules Perform Well on Most Commonly Used Datasets. Machine Learning 11, 63–90 (1993)
53. Honko, P.: Classification of Complex Structured Objects on the base of Similarity Degrees. In: Kryszkiewicz, M., Peters, J.F., Rybinski, H., Skowron, A. (eds.) RSEISP 2007. LNCS (LNAI), vol. 4585, pp. 553–563. Springer, Heidelberg (2007)
54. Honko, P.: Description and Classification of Complex Structured Objects by Applying Similarity Measures. International Journal of Approximate Reasoning (in print, 2008)
55. Honko, P.: Discovery of Meaningful Relationships in Multirelational Data, Ph.D. Thesis, Supervisor: J. Stepaniuk, Białystok University of Technology, Department of Computer Science (in preparation, 2008)
56. Ilczuk, G., Wakulicz-Deja, A.: Data Preparation for Data Mining in Medical Data Sets. In: Peters, J.F., Skowron, A., Düntsch, I., Grzymała-Busse, J.W., Orłowska, E., Polkowski, L. (eds.) Transactions on Rough Sets VI. LNCS, vol. 4374, pp. 83–93. Springer, Heidelberg (2007)
57. Ilczuk, G., Wakulicz-Deja, A.: Selection of Important Attributes for Medical Diagnosis Systems. In: Peters, J.F., Skowron, A., Marek, V.W., Orłowska, E., Słowiński, R., Ziarko, W. (eds.) Transactions on Rough Sets VII. LNCS, vol. 4400, pp. 70–84. Springer, Heidelberg (2007)
58. Ilczuk, G., Wakulicz-Deja, A.: Visualization of Rough Set Decision Rules for Medical Diagnosis Systems. In: An, A., Stefanowski, J., Ramanna, S., Butz, C.J., Pedrycz, W., Wang, G. (eds.) RSFDGrC 2007. LNCS (LNAI), vol. 4482, pp. 371–378. Springer, Heidelberg (2007)
59. Jain, A.K., Murty, M.N., Flynn, P.J.: Data Clustering: a review. ACM Computing Surveys 31(3), 264–323 (1999)
60. Jankowski, A., Peters, J.F., Skowron, A., Stepaniuk, J.: Optimization in Discovery of Compound Granules. Fundamenta Informaticae (2008)
61. Jelonek, J., Krawiec, K., Słowiński, R., Szymaś, J.: Rough Set Reduction of Features for Picture–Based Reasoning. In: Lin, T.Y., Wildberger, A.M. (eds.) Soft Computing: Rough Sets, Fuzzy Logic, Neural Networks, Uncertainty Management, Knowledge Discovery, pp. 89–92. Simulation Councils, Inc., San Diego (1995)

62. Kacprzyk, J.: Linguistic Summaries of Static and Dynamic Data: Computing with Words and Granularity. In: IEEE International Conference on Granular Computing (GrC 2007), pp. 4–5 (2007)
63. Kacprzyk, J., Wilbik, A., Zadrozny, S.: Linguistic Summarization of Time Series Under Different Granulation of Describing Features. In: Kryszkiewicz, M., Peters, J.F., Rybinski, H., Skowron, A. (eds.) RSEISP 2007. LNCS (LNAI), vol. 4585, pp. 230–240. Springer, Heidelberg (2007)
64. Kaelbling, L.P., Littman, M.L., Moore, A.W.: Reinforcement learning: A survey. Journal of Artificial Intelligence Research 4, 227–303 (1996)
65. Kandel, A., Last, M. (eds.): Advances in Fuzzy Logic. Information Sciences An International Journal 177, 2 (2007)
66. Kandulski, M., Marciniec, J., Tukałło, K.: Surgical Wound Infection – Conductive Factors and Their Mutual Dependencies. In: Slowinski, R. (ed.) Intelligent Decision Support - Handbook of Applications and Advances of the Rough Sets Theory, pp. 95–110. Kluwer Academic Publishers, Dordrecht (1992)
67. Kim, D., Bang, S.Y.: A Handwritten Numeral Character Classification Using Tolerant Rough Set. IEEE Transactions on Pattern Analysis and Machine Intelligence 22(9), 923–937 (2000)
68. King, B.: Step-Wise Clustering Procedures. Journal of the American Statistical Association 69, 86–101 (1967)
69. Kloesgen, W., Żytkow, J. (eds.): Handbook of Knowledge Discovery and Data Mining. Oxford University Press, Oxford (2002)
70. Kohonen, T.: Self-Organizing Maps, 2nd edn. Springer, Heidelberg (1997)
71. Koronacki, J., Cwik, J.: Statystyczne systemy uczace sie (a textbook in Polish on statistical learning methodologies), WNT, Warsaw (2005)
72. Krawiec, K., Słowiński, R., Vanderpooten, D.: Learning Decision Rules from Similarity Based Rough Approximations. In: Polkowski, L., Skowron, A. (eds.) Rough Sets in Knowledge Discovery 2. Applications, Case Studies and Software Systems, pp. 37–54. Physica-Verlag, Heidelberg (1998)
73. Kryszkiewicz, M.: Maintenance of Reducts in the Variable Precision Rough Set Model. In: Lin, T.Y., Cercone, N. (eds.) Rough Sets and Data Mining Analysis of Imprecise Data, pp. 355–372. Kluwer Academic Publishers, Dordrecht (1997)
74. Kaufman, L., Rousseeuw, P.J.: Finding Groups in Data: An Introduction to Cluster Analysis. Wiley, Chichester (1990)
75. Kużelewska, U.: Data Mining by Clustering and Information Granulation, Ph.D. Thesis, Supervisor: J. Stepaniuk, Białystok University of Technology, Department of Computer Science (in preparation, 2008)
76. Langley, P., Iba, W.: Average-Case Analysis of a Nearest Neighbor Algorithm. In: Proceedings of the 13th International Joint Conference on Artificial Intelligence, pp. 889–894. Morgan Kaufmann, San Mateo (1993)
77. Leśniewski, S.: Grungzüge eines neuen Systems der Grundlagen der Mathematik. Fundamenta Matemaicae XIV, 1–81 (1929)
78. Luck, M., McBurney, P., Preist, Ch.: Agent Technology: Enabling Next Generation Computing: A Roadmap for Agent Based Computing. In: AgentLink 2003 (2003)
79. Łukasiewicz, J.: Die logischen Grundlagen der Wahrscheinilchkeitsrechnung, Kraków 1913. In: Borkowski, L. (ed.) Jan Łukasiewicz - Selected Works. North Holland, Amsterdam. Polish Scientific Publishers, Warsaw (1970)

80. Li, D., Deogun, J., Spaulding, W., Shuart, B.: Dealing with Missing Data: Algorithms Based on Fuzzy Set and Rough Set Theorie. In: Peters, J.F., Skowron, A. (eds.) Transactions on Rough Sets IV. LNCS, vol. 3700, pp. 37–57. Springer, Heidelberg (2005)

81. Lin, T.Y.: Granular Computing on Binary Relations I Data Mining and Neighborhood Systems. In: Polkowski, L., Skowron, A. (eds.) Rough Sets in Knowledge Discovery 1. Methodology and Applications, pp. 107–121. Physica–Verlag, Heidelberg (1998)

82. Lingras, P., Yao, Y.Y.: Time Complexity of Rough Clustering: GAs versus K-Means. In: Alpigini, J.J., Peters, J.F., Skowron, A., Zhong, N. (eds.) RSCTC 2002. LNCS (LNAI), vol. 2475, pp. 263–270. Springer, Heidelberg (2002)

83. Lingras, P., West, C.: Interval Set Clustering of Web Users with Rough K-Means. Journal of Intelligent Information Systems 23(1), 5–16 (2004)

84. MacQueen, J.: Some Methods for Classification and Analysis of Multivariate Data. In: Le Cam, L.M., Neyman, J. (eds.) Proceedings of the Fifth Berkeley Symposium on Mathematical Statistics and Probability, vol. 1, pp. 281–297. University of California Press, Berkeley (1967)

85. Maji, P., Pal, S.K.: RFCM: A Hybrid Clustering Algorithm Using Rough and Fuzzy Sets. Fundamenta Informaticae 80(4), 475–496 (2007)

86. Martienne, E., Quafafou, M.: Learning Logical Descriptions for Document Understanding: a Rough Sets-Based Approach. In: Polkowski, L., Skowron, A. (eds.) RSCTC 1998. LNCS (LNAI), vol. 1424, pp. 202–209. Springer, Heidelberg (1998)

87. Michalski, R.: A Theory and Methodology of Inductive Learning. In: Michalski, R.S., Carbonell, J.G., Mitchell, T.M. (eds.) Machine Learning, An Artificial Intelligence Approach, pp. 83–134 (1983)

88. Michalski, R.S., Larson, J.B.: Selection of most Representative Training Examples and Incremental Generation of VL1 Hypotheses. Report 867 Department of Computer Science, University of Illinois at Urbana-Champaign (1978)

89. Michie, D., Spiegelhalter, D.J., Taylor, C.C. (eds.): Machine learning, Neural and Statistical Classification. Ellis Horwood, New York (1994)

90. Miikkulainen, R., Bednar, J.A., Choe, Y., Sirosh, J.: Computational Maps in the Visual Cortex. Springer, Hiedelberg (2005)

91. Milton, R.S., Maheswari, V.U., Siromoney, A.: Rough Sets and Relational Learning. In: Peters, J.F., Skowron, A., Grzymała-Busse, J.W., Kostek, B. (eds.) Transactions on Rough Sets I. LNCS, vol. 3100, pp. 321–337. Springer, Heidelberg (2004)

92. Mitra, P., Pal, S.K., Siddiqi, M.A.: Non-Convex Clustering using Expectation Maximization Algorithm with Rough Set Initialization. Pattern Recognition Letters 24(6), 863–873 (2003)

93. Mitra, S.: An Evolutionary Rough Partitive Clustering. Pattern Recognition Letters 25, 1439–1449 (2004)

94. Mrózek, A., Płonka, L.: Rough Sets in Image Analysis. Foundations of Computing Decision Sciences 18(3-4), 259–273 (1993)

95. Nguyen, H.S.: Approximate Boolean Reasoning: Foundations and Applications in Data Mining. In: Peters, J.F., Skowron, A. (eds.) Transactions on Rough Sets V. LNCS, vol. 4100, pp. 344–523. Springer, Heidelberg (2006)

96. Nguyen, H.S., Skowron, A., Stepaniuk, J.: Granular Computing: A Rough Set Approach. An International Journal of Computational Intelligence 17(3), 514–544 (2001)

97. Nguyen, S.H., Bazan, J., Skowron, A., Nguyen, H.S.: Layered learning for concept synthesis. In: Peters, J.F., Skowron, A., Grzymała-Busse, J.W., Kostek, B. (eds.) Transactions on Rough Sets I. LNCS, vol. 3100, pp. 187–208. Springer, Heidelberg (2004)
98. Nicoletti, M.C., Uchoa, J.Q., Baptistini, M.T.Z.: Rough Relation Properties. Int. J. Appl. Math. Comput. Sci. 11(3), 621–635 (2001)
99. Ogiela, M.R., Tadeusiewicz, R.: Modern Computational Intelligence Methods for the Interpretation of Medical Image. Studies in Computational Intelligence, vol. 84. Springer, Heidelberg (2008)
100. Ohrn, A., Komorowski, J., Skowron, A., Synak, P.: The Design and Implementation of a Knowledge Discovery Toolkit Based on Rough Sets - The Rosetta System. In: Polkowski, L., Skowron, A. (eds.) Rough Sets in Knowledge Discovery 1, Methodology and Applications, pp. 376–399. Physica-Verlag, Heidelberg (1998)
101. Orłowska, E.: Information Algebras. In: Alagar, V.S., Nivat, M. (eds.) AMAST 1995. LNCS, vol. 936, pp. 55–65. Springer, Heidelberg (1995)
102. Pal, S.K., Mitra, P.: Pattern Recognition Algorithms for Data Mining: Scalability, Knowledge Discovery, and Soft Granular Computing. Chapman & Hall, Ltd., London (2004)
103. Pal, S.K., Skowron, A. (eds.): Rough-Fuzzy Hybridization A New Trend in Decision Making. Springer, Heidelberg (1999)
104. Pal, S.K., Polkowski, L., Skowron, A. (eds.): Rough–Neural Computing: Techniques for Computing with Words. Springer, Berlin (2004)
105. Pawlak, Z.: Rough Relations. Bulletin of the Polish Academy of Sciences. Technical Sciences 34(9-10), 587–590 (1986)
106. Pawlak, Z.: Rough Sets. In: Theoretical Aspects of Reasoning about Data. Kluwer Academic Publishers, Dordrecht (1991)
107. Pawlak, Z., Skowron, A.: Rough Membership Functions. In: Fedrizzi, M., Kacprzyk, J., Yager, R.R. (eds.) Advances in the Dempster-Shafer Theory of Evidence, pp. 251–271. John Wiley and Sons, New York (1994)
108. Pawlak, Z., Skowron, A.: Rudiments of rough sets. An International Journal of Information Sciences 177(1), 3–27 (2007)
109. Pawlak, Z., Skowron, A.: Rough sets: Some extensions. An International Journal of Information Sciences 177(1), 28–40 (2007)
110. Pawlak, Z., Skowron, A.: Rough sets and Boolean reasoning. An International Journal of Information Sciences 177(1), 41–73 (2007)
111. Pawlak, Z., Słowiński, K., Słowiński, R.: Rough Classification of Patients After Highly Selected Vagotomy for Duodenal Ulcer. Journal of Man–Machine Studies 24, 413–433 (1986)
112. Pedrycz, W. (ed.): Granular Computing. Physica-Verlag, Heidelberg (2001)
113. Pedrycz, W., Skowron, A., Kreinovich, V. (eds.): Handbook of Granular Computing. John Wiley & Sons, New York (2008)
114. Peters, G.: Outliers in Rough k-Means Clustering. In: Pal, S.K., Bandyopadhyay, S., Biswas, S. (eds.) PReMI 2005. LNCS, vol. 3776, pp. 702–707. Springer, Heidelberg (2005)
115. Peters, G., Lampart, M.: A Partitive Rough Clustering Algorithm. In: Greco, S., Hata, Y., Hirano, S., Inuiguchi, M., Miyamoto, S., Nguyen, H.S., Słowiński, R. (eds.) RSCTC 2006. LNCS (LNAI), vol. 4259, pp. 657–666. Springer, Heidelberg (2006)
116. Peters, J.F., Skowron, A., Stepaniuk, J.: Nearness of Objects: Extension of Approximation Space Model. Fundamenta Informaticae 79(3/4), 497–512 (2007)

117. Poggio, T., Smale, S.: The mathematics of learning: Dealing with data. Notices of the AMS 50(5), 537–544 (2003)
118. Polkowski, L.: Rough Sets: Mathematical Foundations. Advances in Soft Computing. Physica-Verlag, Heidelberg (2002)
119. Polkowski, L.: The Paradigm of Granular Rough Computing: Foundations and Applications. In: Zhang, D., Wang, Y., Kinsner, W. (eds.) Proceedings of the Six IEEE International Conference on Cognitive Informatics, ICCI 2007, Lake Tahoe, CA, USA, August 6-8, pp. 145–153. IEEE, Los Alamitos (2007)
120. Polkowski, L., Artiemjew, P.: On Granular Rough Computing with Missing Values. In: Kryszkiewicz, M., Peters, J.F., Rybinski, H., Skowron, A. (eds.) RSEISP 2007. LNCS (LNAI), vol. 4585, pp. 271–279. Springer, Heidelberg (2007)
121. Polkowski, L., Skowron, A.: Rough Mereology. In: Raś, Z.W., Zemankova, M. (eds.) ISMIS 1994. LNCS, vol. 869, pp. 85–94. Springer, Heidelberg (1994)
122. Polkowski, L., Skowron, A.: Rough mereology: A new paradigm for approximate reasoning. Journal of Approximate Reasoning 15(4), 333–365 (1996)
123. Polkowski, L., Skowron, A. (eds.): Rough Sets in Knowledge Discovery 1: Methodology and Applications. Physica-Verlag, Heidelberg (1998)
124. Polkowski, L., Skowron, A. (eds.): Rough Sets in Knowledge Discovery 2: Applications, Case Studies and Software Systems. Physica-Verlag, Heidelberg (1998)
125. Polkowski, L., Skowron, A.: Towards Adaptive Calculus of Granules. In: Proceedings of FUZZ-IEEE 1998 International Conference, Anchorage, Alaska, USA, May 5-9, pp. 111–116 (1998)
126. do Prado, H.A., Engel, P.M., Filho, H.C.: Rough Clustering: An Alternative to Find Meaningful Clusters by Using the Reducts from a Dataset. In: Alpigini, J.J., Peters, J.F., Skowron, A., Zhong, N. (eds.) RSCTC 2002. LNCS (LNAI), vol. 2475, pp. 234–238. Springer, Heidelberg (2002)
127. Rissanen, J.: Modeling by shortes data description. Automatica 14, 465–471 (1978)
128. Rissanen, J.: Minimum-description-length principle. In: Kotz, S., Johnson, N. (eds.) Encyclopedia of Statistical Sciences, pp. 523–527. John Wiley & Sons, New York (1985)
129. Rutkowski, L.: Computational Intelligence, Methods and Techniques. Springer, Heidelberg (2008)
130. Rybinski, H., Kryszkiewicz, M., Protaziuk, G., Jakubowski, A., Delteil, A.: Discovering Synonyms Based on Frequent Termsets. In: Kryszkiewicz, M., Peters, J.F., Rybinski, H., Skowron, A. (eds.) RSEISP 2007. LNCS (LNAI), vol. 4585, pp. 516–525. Springer, Heidelberg (2007)
131. Schalkoff, R.: Pattern Recognition: Statistical, Structural and Neural Approaches. Wiley, Chichester (1992)
132. Sikora, M.: Fuzzy Rules Generation Method for Classification Problems Using Rough Sets and Genetic Algorithms. In: Ślęzak, D., Wang, G., Szczuka, M.S., Düntsch, I., Yao, Y. (eds.) RSFDGrC 2005. LNCS (LNAI), vol. 3641, pp. 383–391. Springer, Heidelberg (2005)
133. Sikora, M., Michalak, M.: NetTRS – Induction and Postprocessing of Decision Rules. In: Greco, S., Hata, Y., Hirano, S., Inuiguchi, M., Miyamoto, S., Nguyen, H.S., Słowiński, R. (eds.) RSCTC 2006. LNCS (LNAI), vol. 4259, pp. 378–387. Springer, Heidelberg (2006)
134. Skowron, A.: Rough sets in KDD (plenary lecture). In: Shi, Z., Faltings, B., Musen, M. (eds.) 16th World Computer Congress (IFIP 2000): Proceedings of Conference on Intelligent Information Processing (IIP 2000), pp. 1–17. Publishing House of Electronic Industry, Beijing (2000)

135. Skowron, A.: Approximate reasoning in distributed environments. In: Zhong, N., Liu, J. (eds.) Intelligent Technologies for Information Analysis, pp. 433–474. Springer, Heidelberg (2004)
136. Skowron, A., Polkowski, L.: Synthesis of Decision Systems from Data Tables. In: Lin, T.Y., Cercone, N. (eds.) Rough Sets and Data Mining Analysis of Imprecise Data, pp. 259–299. Kluwer Academic Publishers, Dordrecht (1997)
137. Skowron, A., Rauszer, C.: The Discernibility Matrices and Functions in Information Systems. In: Słowiński, R. (ed.) Intelligent Decision Support. Handbook of Applications and Advances of Rough Sets Theory, pp. 331–362. Kluwer Academic Publishers, Dordrecht (1992)
138. Skowron, A., Stepaniuk, J.: Towards an Approximation Theory of Discrete Problems. Fundamenta Informaticae 15(2), 187–208 (1991)
139. Skowron, A., Stepaniuk, J.: Searching for Classifiers. In: De Glas, M., Gabbay, D. (eds.) Proceedings of the First World Conference on the Fundamentals of Artificial Intelligence (WOCFAI 1991), Angkor, Paris, July 1-5, pp. 447–460 (1991)
140. Skowron, A., Stepaniuk, J.: Intelligent Systems Based on Rough Set Approach. Foundations of Computing and Decision Sciences 18(3-4), 343–360 (1993)
141. Skowron, A., Stepaniuk, J.: Approximations of Relations. In: Ziarko, W. (ed.) Rough Sets, Fuzzy Sets and Knowledge Discovery, pp. 161–166. Springer, London, Berlin (1994); see also: Institute of Computer Science, Warsaw University of Technology, ICS Research Report 20/94 (1994)
142. Skowron, A., Stepaniuk, J.: Generalized Approximation Spaces. In: Proceedings of the Third International Workshop on Rough Sets and Soft Computing, San Jose, November 10-12, pp. 156–163 (1994)
143. Skowron, A., Stepaniuk, J.: Generalized Approximation Spaces. In: Lin, T.Y., Wildberger, A.M. (eds.) Soft Computing, Simulation Councils, San Diegon, pp. 18–21 (1995); see also: Institute of Computer Science, Warsaw University of Technology, ICS Research Report 41/94 (1994)
144. Skowron, A., Stepaniuk, J.: Decision Rules Based on Discernibility Matrices and Decision Matrices. In: Lin, T.Y., Wildberger, A.M. (eds.) Soft Computing, Simulation Councils, San Diego, pp. 6–9 (1995); see also Institute of Computer Science, Warsaw University of Technology, ICS Research Report 40/94 (1994)
145. Skowron, A., Stepaniuk, J.: Tolerance approximation spaces. Fundamenta Informaticae 27, 245–253 (1996)
146. Skowron, A., Stepaniuk, J.: Information Reduction Based on Constructive Neighborhood Systems. In: Wang, P.P. (ed.) Proceedings of the Fifth International Workshop on Rough Sets and Soft Computing (RSSC 1997) at Third Annual Joint Conference on Information Sciences (JCIS 1997), Rough Set & Computer Science, Duke University, Durham, NC, USA, March 1–5, vol. 3, pp. 158–160 (1997)
147. Skowron, A., Stepaniuk, J.: Constructive Information Granules. In: Proceedings of the 15th IMACS World Congress on Scientific Computation, Modelling and Applied Mathematics, Artificial Intelligence and Computer Science, Berlin, Germany, August 24-29, vol. 4, pp. 625–630 (1997)
148. Skowron, A., Stepaniuk, J.: Information Granules and Approximation Spaces. In: Proceedings of Seventh International Conference on Information Processing and Management of Uncertainty in Knowledge-Based Systems, Paris, France, July 6-10, pp. 354–361 (1998)
149. Skowron, A., Stepaniuk, J.: Towards Discovery of Information Granules. In: Żytkow, J.M., Rauch, J. (eds.) PKDD 1999. LNCS (LNAI), vol. 1704, pp. 542–547. Springer, Heidelberg (1999)

150. Skowron, A., Stepaniuk, J.: Information Granules in Distributed Environment. In: Zhong, N., Skowron, A., Ohsuga, S. (eds.) RSFDGrC 1999. LNCS (LNAI), vol. 1711, pp. 357–365. Springer, Heidelberg (1999)

151. Skowron, A., Stepaniuk, J.: Information Granule Decomposition. Fundamenta Informaticae 47(3-4), 337–350 (2001)

152. Skowron, A., Stepaniuk, J.: Information granules: Towards foundations of granular computing. International Journal of Intelligent Systems 16(1), 57–86 (2001)

153. Skowron, A., Stepaniuk, J.: Information Granules: Towards Foundations for Spatial and Temporal Reasoning. Proceedings of the Indian National Science Academy 67A(2), 315–325 (2001)

154. Skowron A., Stepaniuk J.: Information granules and rough-neural computing. In [103], 43–84

155. Skowron, A., Stepaniuk, J.: Constrained sums of information systems. In: Tsumoto, S., Słowiński, R., Komorowski, J., Grzymała-Busse, J.W. (eds.) RSCTC 2004. LNCS (LNAI), vol. 3066, pp. 300–309. Springer, Heidelberg (2004)

156. Skowron, A., Stepaniuk, J., Peters, J.F.: Towards Discovery of Relevant Patterns from Parameterized Schemes of Information Granule Construction. In: Inuiguchi, M., Hirano, S., Tsumoto, S. (eds.) Rough Set Theory and Granular Computing, pp. 97–108 (2003)

157. Skowron, A., Stepaniuk, J., Peters, J.F.: Rough Sets and Infomorphisms: Towards Approximation of Relations in Distributed Environments. Fundamenta Informaticae 54(1-2), 263–277 (2003)

158. Skowron, A., Stepaniuk, J., Peters, J.F., Swiniarski, R.: Calculi of Approximation Spaces. Fundamenta Informaticae 72(1-3), 363–378 (2006)

159. Skowron, A., Stepaniuk, J.: Modeling of High Quality Granules. In: Kryszkiewicz, M., Peters, J.F., Rybinski, H., Skowron, A. (eds.) RSEISP 2007. LNCS (LNAI), vol. 4585, pp. 300–309. Springer, Heidelberg (2007)

160. Skowron, A., Synak, P.: Complex patterns. Fundamenta Informaticae 60(1-4), 351–366 (2004)

161. Skowron, A., Swiniarski, R., Synak, P.: Approximation spaces and information granulation. In: Peters, J.F., Skowron, A. (eds.) Transactions on Rough Sets III. LNCS, vol. 3400, pp. 175–189. Springer, Heidelberg (2005)

162. Słowiński, R.: Strict and Weak Indiscernibility of Objects Described by Quantitative Attributes with Overlapping Norms. Foundations of Computing and Decision Sciences 18, 361–369 (1993)

163. Słowiński, R., Greco, S.: Measuring Attractiveness of Rules from the Viewpoint of Knowledge Representation. In: Szczepaniak, P.S., Kacprzyk, J., Niewiadomski, A. (eds.) AWIC 2005. LNCS (LNAI), vol. 3528, pp. 11–22. Springer, Heidelberg (2005)

164. Słowiński, R., Greco, S., Matarazzo, B.: Dominance-Based Rough Set Approach to Reasoning About Ordinal Data. In: Kryszkiewicz, M., Peters, J.F., Rybinski, H., Skowron, A. (eds.) RSEISP 2007. LNCS (LNAI), vol. 4585, pp. 5–11. Springer, Heidelberg (2007)

165. Słowiński, K., Słowiński, R., Stefanowski, J.: Rough Sets Approach to Analysis of Data from Peritoneal Lavage in Acute Pancreatitis. Medical Informatics 13(3), 143–159 (1988)

166. Słowiński, R., Stefanowski, J.: Software Implementation of the Rough Set Theory. In: Polkowski, L., Skowron, A. (eds.) Rough Sets in Knowledge Discovery 2. Applications, Case Studies and Software Systems, pp. 581–586. Physica-Verlag, Heidelberg (1998)

167. Słowiński, R., Szczech, I., Urbanowicz, M., Greco, S.: Mining Association Rules with Respect to Support and Anti-support-Experimental Results. In: Kryszkiewicz, M., Peters, J.F., Rybinski, H., Skowron, A. (eds.) RSEISP 2007. LNCS (LNAI), vol. 4585, pp. 534–542. Springer, Heidelberg (2007)

168. Sneath, P.H.A., Sokal, R.R.: Numerical Taxonomy. Freeman, San Francisco (1973)

169. Staab, S., Studer, R. (eds.): Handbook on Ontologies. International Handbooks on Information Systems. Springer, Heidelberg (2004)

170. Stepaniuk, J.: Elementary Approximation Theory. Bulletin of the Polish Academy of Sciences Tech. 38(1-12), 121–128 (1990)

171. Stepaniuk, J.: Approximation Logic of Programs. Bulletin of the Polish Academy of Sciences Tech. 38(1-12), 129–138 (1990)

172. Stepaniuk, J.: Applications of Finite Models Properties in Approximation and Algorithmic Logics. Fundamenta Informaticae 14(1), 91–108 (1991)

173. Stepaniuk, J.: Methods of Approximate Reasoning for Discrete Problems. Ph.D. Dissertation, Warsaw University (1992)

174. Stepaniuk, J.: Decision Rules for Consistent Decision Tables. In: Proceedings of the Polish–English Meeting on Information Systems, Bialystok, Poland, September 22, pp. 76–86 (1993)

175. Stepaniuk, J.: Decision Rules for Decision Tables. Bulletin of the Polish Academy of Sciences Tech. 42(3), 457–469 (1994)

176. Stepaniuk, J.: Discernibility and Decision Matrices (in Polish). In: Kulikowski, R., Bogdan, L. (eds.) Wspomaganie Decyzji, Systemy Eksperckie, Institute of System Analysis PAS, Warsaw, Poland, pp. 440–443 (1995)

177. Stepaniuk, J.: Properties and Applications of Rough Relations. In: Proceedings of the Fifth International Workshop on Intelligent Information Systems, Deblin, Poland, June 2-5, pp. 136–141. Institute of Computer Science, Polish Academy of Sciences, Warsaw (1996); see also Institute of Computer Science, Warsaw University of Technology, ICS Research Report 26/96 (1996)

178. Stepaniuk, J.: Similarity Based Rough Sets and Learning. In: Tsumoto, S., Kobayashi, S., Yokomori, T., Tanaka, H. (eds.) Proceedings of the Fourth International Workshop on Rough Sets, Fuzzy Sets and Machine Discovery (RSFD 1996), Tokyo, November 6-8, pp. 18–22 (1996)

179. Stepaniuk, J.: Attribute Discovery and Rough Sets. In: Principles of Data Mining and Knowledge Discovery, First European Symposium, PKDD 1997, Trondheim, Norway. LNCS (LNAI), vol. 1263, pp. 145–155. Springer, Heidelberg (June 1997)

180. Stepaniuk, J.: Rough Sets, First Order Logic and Attribute Construction. In: Proceedings of the Sixth International Conference, Information Processing and Management of Uncertainty in Knowledge–Based Systems (IPMU 1996), Granada, Spain, July 1-5, vol. 2, pp. 887–890 (1996)

181. Stepaniuk, J.: Rough Sets Similarity Based Learning. In: Proceedings of the Fifth European Congress on Intelligent Techniques and Soft Computing, Aachen, Germany, September 8-12, pp. 1634–1638. Verlag Mainz (1997)

182. Stepaniuk, J.: Conflict Analysis and Groups of Agents. In: Raś, Z.W., Skowron, A. (eds.) ISMIS 1997. LNCS, vol. 1325, pp. 174–185. Springer, Heidelberg (1997)

183. Stepaniuk, J.: Approximation Spaces in Extensions of Rough Set Theory. In: Polkowski, L., Skowron, A. (eds.) RSCTC 1998. LNCS (LNAI), vol. 1424, pp. 290–297. Springer, Heidelberg (1998)

184. Stepaniuk, J.: Optimizations of Rough Set Model. Fundamenta Informaticae 36(2-3), 265–283 (1998)

185. Stepaniuk, J.: Rough relations and logics. In: Polkowski, L., Skowron, A. (eds.) Rough Sets in Knowledge Discovery 1. Methodology and Applications, pp. 248–260. Physica Verlag, Heidelberg (1998)

186. Stepaniuk, J.: Approximation Spaces, Reducts and Representatives. In: Polkowski, L., Skowron, A. (eds.) Rough Sets in Knowledge Discovery 2. Applications, Case Studies and Software Systems, pp. 109–126. Physica-Verlag, Heidelberg (1998)

187. Stepaniuk, J.: Rough Set Data Mining of Diabetes Data. In: Raś, Z.W., Skowron, A. (eds.) ISMIS 1999. LNCS, vol. 1609, pp. 457–465. Springer, Heidelberg (1999)

188. Stepaniuk, J.: Rough Sets and Relational Learning. In: Proceedings of the Seventh European Congress on Intelligent Techniques and Soft Computing, Aachen, Germany, September 13-16, 6 pages. Verlag Mainz (1999) (CD-ROM)

189. Stepaniuk, J.: Knowledge discovery by application of rough set models. In: Polkowski, L., Tsumoto, S., Lin, T.Y. (eds.) Rough Set Methods and Applications. New Developments in Knowledge Discovery in Information Systems, pp. 137–233. Physica–Verlag, Heidelberg (2000)

190. Stepaniuk, J.: Tolerance Information Granules. In: Dunin-Kȩplicz, B., Jankowski, A., Skowron, A., Szczuka, M. (eds.) Monitoring, Security and Rescue Techniques in Multiagent Systems, pp. 305–316. Springer, Heidelberg (2005)

191. Stepaniuk, J.: Relational Data and Rough Sets. Fundamenta Informaticae 79(3/4), 525–539 (2007)

192. Stepaniuk, J.: Approximation Spaces in Multi Relational Knowledge Discovery. In: Peters, J.F., Skowron, A., Düntsch, I., Grzymała-Busse, J.W., Orłowska, E., Polkowski, L. (eds.) Transactions on Rough Sets VI. LNCS, vol. 4374, pp. 351–365. Springer, Heidelberg (2007)

193. Stepaniuk, J., Bazan, J., Skowron, A.: Modelling complex patterns by information systems. Fundamenta Informaticae 67(1-3), 203–217 (2005)

194. Stepaniuk, J., Góralczuk, L.: An Algorithm Generating First Order Rules Based on Rough Set Methods. In: Stepaniuk, J. (ed.) Zeszyty Naukowe Politechniki Białostockiej Informatyka, vol. 1, pp. 235–250 (2002) [in Polish]

195. Stepaniuk, J., Honko, P.: Learning First–Order Rules: A Rough Set Approach. Fundamenta Informaticae 61(2), 139–157 (2004)

196. Stepaniuk, J., Krętowski, M.: Decision System Based on Tolerance Rough Sets. In: Proceedings of the Fourth International Workshop on Intelligent Information Systems, Augustow, Poland, June 5-9, pp. 62–73. Institute of Computer Science, Polish Academy of Sciences, Warsaw (1995); see also Institute of Computer Science, Warsaw University of Technology, ICS Research Report 36/95 (1995)

197. Stepaniuk, J., Kużelewska, U.: Granulation using Clustering: A Medical Case Study. In: Proceedings of CS&P 2007, vol. 2, pp. 509–520 (2007)

198. Stepaniuk, J., Maj, M.: Data Transformation and Rough Sets. In: Żytkow, J.M. (ed.) PKDD 1998. LNCS, vol. 1510, pp. 441–449. Springer, Heidelberg (1998)

199. Stepaniuk, J., Tyszkiewicz, J.: Probabilistic Properties of Approximation Problems. Bulletin of the Polish Academy of Sciences Tech. 39(3), 535–555 (1991)

200. Stone, P.: Layered Learning in Multi-Agent Systems: A Winning Approach to Robotic Soccer. MIT Press, Cambridge (2000)

201. Sutton, R.S., Barto, A.G.: Reinforcement Learning: An Introduction. MIT Press, Cambridge (1998)

202. Ślȩzak, D.: Approximate entropy reducts. Fundamenta Informaticae 53(3-4), 365–390 (2002)

203. Torgo, L.: Controlled Redundancy in Incremental Rule Learning. In: Brazdil, P.B. (ed.) ECML 1993. LNCS, vol. 667, pp. 185–195. Springer, Heidelberg (1993)

204. Tsumoto, S.: Extraction of Experts Decision Process from Clinical Databases Using Rough Set Model. In: Komorowski, J., Żytkow, J.M. (eds.) PKDD 1997. LNCS, vol. 1263, pp. 58–67. Springer, Heidelberg (1997)

205. Tsumoto, S.: Formalization and Induction of Medical Expert System Rules Based on Rough Set Theory. In: Polkowski, L., Skowron, A. (eds.) Rough Sets in Knowledge Discovery 2. Applications, Case Studies and Software Systems, pp. 307–323. Physica-Verlag, Heidelberg (1998)

206. Urmson, C., et al.: High speed navigation of unrehearsed terrain: Red team technology for grand challenge 2004. Technical Report CMU-RI-TR-04-37, Robotics Institute, Carnegie Mellon University, Pittsburgh, PA (June 2004)

207. Vakarelov, D.: Rough Polyadic Modal Logics. Journal of Applied Non-Classical Logics 1(1), 9–36 (1991)

208. Vapnik, V.: Statistical Learning Theory. Wiley, New York (1998)

209. Wakulicz–Deja, A., Paszek, P.: Diagnose Progressive Encephalopathy Applying the Rough Set Theory. International Journal of Medical Informatics 46, 119–127 (1997)

210. Wang, W., et al.: Sting: A Statistical Information Grid Approach to Spatial Data Mining. In: Proceedings of International Conference on Very Large Data Bases, pp. 186–195. Morgan - Kaufmann, Athens (1997)

211. Ward, J.H.: Hierarchical Grouping to Optimize an Objective Function. Journal of American Statistical Association 58(301), 236–244 (1963)

212. Wierzbicki, J.A.: Rough Sets in Case-Based Reasoning, Ph.D. Thesis, Supervisor: J. Stepaniuk, Białystok University of Technology, Department of Computer Science (2004)

213. Wierzchoń, S.T., Kużelewska, U.: Evaluation of Clusters Quality in Artificial Immune Clustering System - SArIS. In: Biometrics, Computer Security Systems and Artificial Intelligence Applications, pp. 323–331. Springer, Heidelberg (2006)

214. Wilson, D.A., Martinez, T.R.: Improved Heterogeneous Distance Functions. Journal of Artificial Intelligence Research 6, 1–34 (1997)

215. Witten, I.H., Frank, E.: Data Mining: Practical Machine Learning Tools and Techniques, 2nd edn. Morgan - Kaufmann, San Francisco (2005)

216. Wojna, A.: Analogy-based Reasoning in Classifier Construction. In: Peters, J.F., Skowron, A. (eds.) Transactions on Rough Sets IV. LNCS, vol. 3700, pp. 277–374. Springer, Heidelberg (2005)

217. Wu, X., Kumar, V., Quinlan, J.R., Ghosh, J., Yang, Q., Motoda, H., McLachlan, G.J., Ng, A., Liu, B., Yu, P.S., Zhou, Z.H., Steinbach, M., Hand, D.J., Steinberg, D.: Top 10 Algorithms in Data Mining. Knowledge and Information Systems 14(1), 1–37 (2008)

218. Yang, Q., Wu, X.: 10 Challenging Problems in Data Mining Research. International Journal of Information Technology & Decision Making 5(4), 597–604 (2006)

219. Yao, Y.Y., Zhong, N.: An Analysis of Quantitative Measures Associated with Rules. In: Zhong, N., Zhou, L. (eds.) PAKDD 1999. LNCS (LNAI), vol. 1574, pp. 479–488. Springer, Heidelberg (1999)

220. Zadeh, L.A.: Outline of a new approach to the analysis of complex system and decision processes. IEEE Transactions on Systems, Man, and Cybernetics SMC3, 28–44 (1973)

221. Zadeh L.A.: The concept of a linguistic variable and its application to approximate reasoning, Part I: Information Sciences 8, 199–249 (1975); Part II: Information Sciences 8, 301–357 (1975); Part III: Information Sciences 9, 43–80 (1975)

222. Zadeh, L.A.: Fuzzy Sets and Information Granularity. In: Gupta, M., Ragade, R., Yager, R. (eds.) Advances in Fuzzy Set Theory and Applications, pp. 3–18. North-Holland Publishing Co., Amsterdam (1979)
223. Zadeh, L.A.: Fuzzy Logic = Computing with Words. IEEE Trans. on Fuzzy Systems 4, 103–111 (1996)
224. Zadeh, L.A.: Toward a Theory of Fuzzy Information Granulation and Its Certainty in Human Reasoning and Fuzzy Logic. Fuzzy Sets and Systems 90, 111–127 (1997)
225. Zadeh, L.A.: From computing with numbers to computing with words – From manipulation of measurements to manipulation of perceptions. IEEE Transactions on Circuits and Systems 45, 105–119 (1999)
226. Zadeh, L.A.: A new direction in AI: Toward a computational theory of perceptions. AI Magazine 22(1), 73–84 (2001)
227. Zadeh, L.A., Kacprzyk, J. (eds.): Computing with Words in Information/Intelligent Systems 1. Foundations. Physica-Verlag, Heidelberg (1999)
228. Zadeh, L.A., Kacprzyk, J. (eds.): Computing with Words in Information/Intelligent Systems 2. Applications. Physica-Verlag, Heidelberg (1999)
229. Ziarko, W.: Variable precision rough set model. Journal of Computer and System Sciences 46, 39–59 (1993)
230. Ziarko, W., Shan, N.: KDD–R: A Comprehensive System for Knowledge Discovery in Databases Using Rough Sets. In: Proceedings of the Third International Workshop on Rough Sets and Soft Computing, San Jose, November 10-12, pp. 164–173 (1994)

A Further Readings

A.1 Books

Cios, K., Pedrycz, W., Swiniarski, R.: Data mining methods for knowledge discovery. Kluwer, Norwell (1998)

Demri, S., Orłowska, E.: Incomplete Information: Structure, Inference, Complexity. In: Monographs in Theoretical Computer Science. Springer, Heidelberg (2002)

Doherty, P., Łukaszewicz, W., Skowron, A., Szałas, A.: Knowledge Engineering: A Rough Sets Approach. Springer Physica-Verlag, Berlin (2006)

Dunin-Kęplicz, B., Jankowski, A., Skowron, A., Szczuka, M.: Monitoring, Security, and Rescue Tasks in Multiagent Systems (MSRAS 2004). Advances in Soft Computing. Springer, Heidelberg (2005)

Düntsch, I., Gediga, G.: Rough set data analysis: A road to non-invasive knowledge discovery. Methodos Publishers, Bangor (2000)

Grzymała-Busse, J.W.: Managing Uncertainty in Expert Systems. Kluwer Academic Publishers, Norwell (1990)

Inuiguchi, M., Hirano, S., Tsumoto, S. (eds.): Rough Set Theory and Granular Computing. Studies in Fuzziness and Soft Computing, vol. 125. Springer, Heidelberg (2003)

Kostek, B.: Soft Computing in Acoustics, Applications of Neural Networks, Fuzzy Logic and Rough Sets to Physical Acoustics. Studies in Fuzziness and Soft Computing, vol. 31. Physica-Verlag, Heidelberg (1999)

Kostek, B.: Perception-Based Data Processing in Acoustics. In: Applications to Music Information Retrieval and Psychophysiology of Hearing. Studies in Computational Intelligence, vol. 3. Springer, Heidelberg (2005)

Lin, T.Y., Yao, Y.Y., Zadeh, L.A. (eds.): Data Mining, Rough Sets and Granular Computing. Studies in Fuzziness and Soft Computing. Physica-Verlag, Heidelberg (2002)

Lin, T.Y., Cercone, N. (eds.): Rough Sets and Data Mining - Analysis of Imperfect Data. Kluwer Academic Publishers, Boston (1997)

Mitra, S., Acharya, T.: Data mining. In: Multimedia, Soft Computing, and Bioinformatics. John Wiley & Sons, New York (2003)

Munakata, T. (ed.): Fundamentals of the New Artificial Intelligence: Beyond Traditional Paradigms. Graduate Texts in Computer Science, vol. 10. Springer, New York (1998)

J. Stepaniuk: Rough - Gran. Comput. in Knowl. Dis. & Data Min., SCI 152, pp. 151–155, 2008.
springerlink.com   © Springer-Verlag Berlin Heidelberg 2008

Orłowska, E. (ed.): Incomplete Information: Rough Set Analysis. Studies in Fuzziness and Soft Computing, vol. 13. Physica-Verlag, Heidelberg (1998)

Pal, S.K., Polkowski, L., Skowron, A. (eds.): Rough-Neural Computing: Techniques for Computing with Words. Cognitive Technologies. Springer, Heidelberg (2004)

Pal, S.K., Skowron, A. (eds.): Rough Fuzzy Hybridization: A New Trend in Decision-Making. Springer, Singapore (1999)

Polkowski, L., Lin, T.Y., Tsumoto, S. (eds.): Rough Set Methods and Applications: New Developments in Knowledge Discovery in Information Systems. Studies in Fuzziness and Soft Computing, vol. 56. Springer, Heidelberg (2000)

Polkowski, L., Skowron, A. (eds.): Rough Sets in Knowledge Discovery 1: Methodology and Applications. Studies in Fuzziness and Soft Computing, vol. 18. Physica-Verlag, Heidelberg (1998)

Polkowski, L., Skowron, A. (eds.): Rough Sets in Knowledge Discovery 2: Applications, Case Studies and Software Systems. Studies in Fuzziness and Soft Computing, vol. 19. Physica-Verlag, Heidelberg (1998)

Słowiński, R. (ed.): Intelligent Decision Support - Handbook of Applications and Advances of the Rough Sets Theory, System Theory, Knowledge Engineering and Problem Solving, vol. 11. Kluwer Academic Publishers, Dordrecht (1992)

Zhong, N., Liu, J. (eds.): Intelligent Technologies for Information Analysis. Springer, Heidelberg (2004)

A.2   Transactions on Rough Sets

Peters, J.F., Skowron, A., Grzymała-Busse, J.W., Kostek, B.z., Świniarski, R.W., Szczuka, M.S. (eds.): Transactions on Rough Sets I. LNCS, vol. 3100. Springer, Heidelberg (2004)

Peters, J.F., Skowron, A., Dubois, D., Grzymała-Busse, J.W., Inuiguchi, M., Polkowski, L. (eds.): Transactions on Rough Sets II. LNCS, vol. 3135. Springer, Heidelberg (2005)

Peters, J.F., Skowron, A., van Albada, D. (eds.): Transactions on Rough Sets III. LNCS, vol. 3400. Springer, Heidelberg (2005)

Peters, J.F., Skowron, A. (eds.): Transactions on Rough Sets IV. LNCS, vol. 3700. Springer, Heidelberg (2005)

Peters, J.F., Skowron, A. (eds.): Transactions on Rough Sets V. LNCS, vol. 4100. Springer, Heidelberg (2006)

Peters, J.F., Skowron, A., Düntsch, I., Grzymała-Busse, J.W., Orłowska, E., Polkowski, L. (eds.): Transactions on Rough Sets VI. LNCS, vol. 4374. Springer, Heidelberg (2007)

Peters, J.F., Skowron, A., Marek, V.W., Orłowska, E., Słowiński, R., Ziarko, W. (eds.): Transactions on Rough Sets VII. LNCS, vol. 4400. Springer, Heidelberg (2007)

A.3   Special Issues of Journals

Cercone, N., Skowron, A., Zhong, N. (eds.): Special issue, Computational Intelligence: An International Journal 17(3) (2001)

Lin, T.Y. (ed.): Special issue, Journal of the Intelligent Automation and Soft Computing 2(2) (1996)

Peters, J.F., Skowron, A. (eds.): Special issue on a rough set approach to reasoning about data. International Journal of Intelligent Systems 16(1) (2001)

Pal, S.K., Pedrycz, W., Skowron, A., Swiniarski, R.(eds.): Special volume: Rough-neuro computing. Neurocomputing 36 (2001)

Skowron, A., Pal, S.K. (eds.): Special volume: Rough sets, pattern recognition and data mining. Pattern Recognition Letters 24(6) (2003)

Słowiński, R., Stefanowski, J. (eds.): Special issue: Proceedings of the First International Workshop on Rough Sets: State of the Art and Perspectives, Kiekrz, Poznań, Poland, September 2-4 (1992); Foundations of Computing and Decision Sciences 18(3-4) (1993)

Ziarko, W. (ed.): Special issue, Computational Intelligence: An International Journal 11(2) (1995)

Ziarko, W. (ed.): Special issue, Fundamenta Informaticae 27(2-3) (1996)

A.4   Proceedings of International Conferences

Alpigini, J.J., Peters, J.F., Skowron, A., Zhong, N. (eds.): RSCTC 2002. LNCS (LNAI), vol. 2475. Springer, Heidelberg (2002)

An, A., Stefanowski, J., Ramanna, S., Butz, C.J., Pedrycz, W., Wang, G. (eds.): RSFD-GrC 2007. LNCS (LNAI), vol. 4482. Springer, Heidelberg (2007)

Greco, S., Hata, Y., Hirano, S., Inuiguchi, M., Miyamoto, S., Nguyen, H.S., Słowiński, R. (eds.): RSCTC 2006. LNCS (LNAI), vol. 4259. Springer, Heidelberg (2006)

Hirano, S., Inuiguchi, M., Tsumoto, S. (eds.): Proceedings of International Workshop on Rough Set Theory and Granular Computing (RSTGC 2001), Matsue, Shimane, Japan, May 20-22 (2001); Bulletin of the International Rough Set Society 5(1-2) (2001)

Kryszkiewicz, M., Peters, J.F., Rybinski, H., Skowron, A. (eds.): RSEISP 2007. LNCS (LNAI), vol. 4585. Springer, Heidelberg (2007)

Lin, T.Y., Wildberger, A.M. (eds.): Soft Computing: Rough Sets, Fuzzy Logic, Neural Networks, Uncertainty Management, Knowledge Discovery. Simulation Councils, Inc., San Diego (1995)

Polkowski, L., Skowron, A. (eds.): RSCTC 1998. LNCS (LNAI), vol. 1424. Springer, Heidelberg (1998)

Skowron, A. (ed.): SCT 1984. LNCS, vol. 208. Springer, Heidelberg (1985)

Skowron, A., Szczuka, M. (eds.): Proceedings of the Workshop on Rough Sets in Knowledge Discovery and Soft Computing at (ETAPS 2003), Elsevier, Amsterdam, Netherlands, April 12-13 (2003); Electronic Notes in Computer Science 82(4) (2003), http://www.elsevier.nl/locate/entcs/volume82.html

Ślęzak, D., Wang, G., Szczuka, M.S., Düntsch, I., Yao, Y. (eds.): RSFDGrC 2005. LNCS (LNAI), vol. 3641. Springer, Heidelberg (2005)

Ślęzak, D., Yao, J.T., Peters, J.F., Ziarko, W., Hu, X. (eds.): RSFDGrC 2005. LNCS (LNAI), vol. 3642. Springer, Heidelberg (2005)

Terano, T., Nishida, T., Namatame, A., Tsumoto, S., Ohsawa, Y., Washio, T. (eds.): JSAI-WS 2001. LNCS (LNAI), vol. 2253. Springer, Heidelberg (2001)

Tsumoto, S., Kobayashi, S., Yokomori, T., Tanaka, H., Nakamura, A. (eds.): Proceedings of the Fourth Internal Workshop on Rough Sets, Fuzzy Sets and Machine Discovery, University of Tokyo, Japan, November 6-8. The University of Tokyo, Tokyo (1996)

Tsumoto, S., Słowiński, R., Komorowski, J., Grzymała-Busse, J. (eds.): RSCTC 2004. LNCS (LNAI), vol. 3066. Springer, Heidelberg (2004)

Yao, J.T., Lingras, P., Wu, W.-Z., Szczuka, M.S., Cercone, N.J., Ślęzak, D. (eds.): RSKT 2007. LNCS (LNAI), vol. 4481. Springer, Heidelberg (2007)

Wang, G., Liu, Q., Yao, Y., Skowron, A. (eds.): RSFDGrC 2003. LNCS (LNAI), vol. 2639. Springer, Heidelberg (2003)

Zhong, N., Skowron, A., Ohsuga, S. (eds.): RSFDGrC 1999. LNCS (LNAI), vol. 1711. Springer, Heidelberg (1999)

Ziarko, W.: Rough Sets, Fuzzy Sets and Knowledge Discovery: Proceedings of the Second International Workshop on Rough Sets and Knowledge Discovery (RSKD 1993). Workshops in Computing, Banff, Alberta, Canada, October 12–15. Springer–Verlag & British Computer Society, London, Berlin (1994)

Ziarko, W., Yao, Y. (eds.): RSCTC 2000. LNCS (LNAI), vol. 2005. Springer, Heidelberg (2001)

A.5   Selected Web Resources

International Rough Set Society (IRSS) is a non-profit organisation intended as a forum for contacts and exchange of information between members of scientific community whos' research is related to the rough set theory, `http://roughsets.home.pl/www/`

RSES (Rough Set Exploration System) is a toolkit for analysis of table data. It is based on methods and algorithms coming from the area of rough sets, `http://logic.mimuw.edu.pl/~rses/`

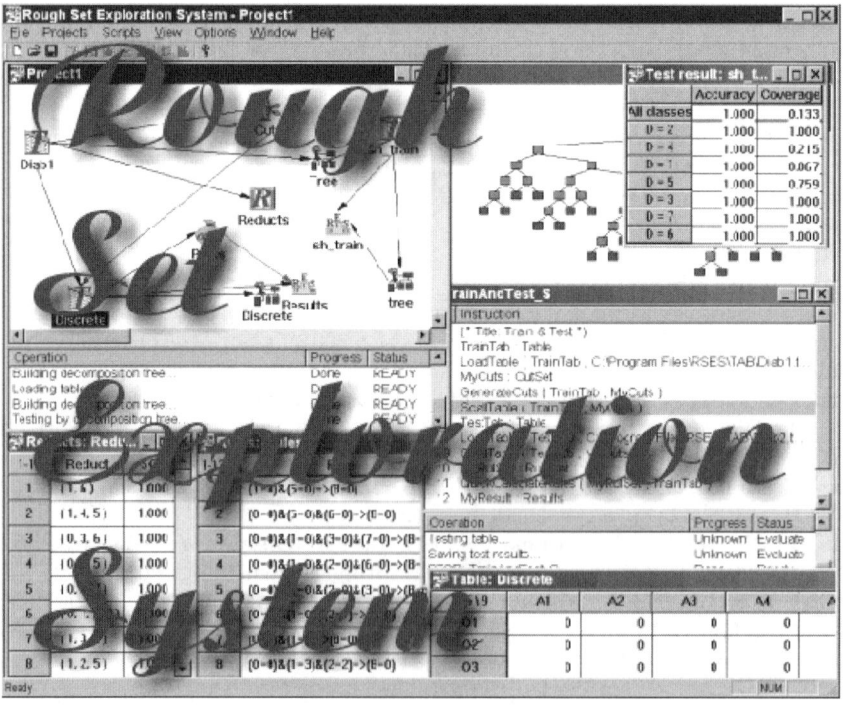

Fig. A.1. Rough Set Exploration System

ROSE (Rough Sets Data Explorer) is a software implementing basic elements of the rough set theory and rule discovery techniques,
`http://idss.cs.put.poznan.pl/site/rose.html`

ACM Special Interest Group on Knowledge Discovery and Data Mining – a Knowledge Discovery and Data Mining Society under the umbrella of ACM,
`http://www.sigkdd.org/`

KDnuggets.com (KD stands for Knowledge Discovery) is the source of information on Data Mining, Web Mining, Knowledge Discovery, and Decision Support Topics, including News, Software, Solutions, Companies, Jobs, Courses, Meetings, and Publications, `http://www.kdnuggets.com/`

Index